W9-BRO-306

Kelley Wingate
Math Practice
Fifth Grade

Credits
Content Editor: Elise Craver
Copy Editor: Angela Triplett

Visit *carsondellosa.com* for correlations to Common Core, state, national, and Canadian provincial standards.

Carson-Dellosa Publishing, LLC
PO Box 35665
Greensboro, NC 27425 USA
carsondellosa.com

ISBN 978-1-4838-0503-0
06-072171151

Table of Contents

Introduction

Competency in basic math skills creates a foundation for the successful use of math principles in the real world. Practicing math skills—in the areas of operations, algebra, place value, fractions, measurement, and geometry—is the best way to improve at them.

This book was developed to help students practice and master basic mathematical concepts. The practice pages can be used first to assess proficiency and later as basic skill practice. The extra practice will help students advance to more challenging math work with confidence. Help students catch up, stay up, and move ahead.

Common Core State Standards (CCSS) Alignment

This book supports standards-based instruction and is aligned to the CCSS. The standards are listed at the top of each page for easy reference. To help you meet instructional, remediation, and individualization goals, consult the Common Core State Standards alignment chart on page 4.

Leveled Activities

Instructional levels in this book vary. Each area of the book offers multilevel math activities so that learning can progress naturally. There are three levels, signified by one, two, or three dots at the bottom of the page:

- Level I: These activities will offer the most support.
- Level II: Some supportive measures are built in.
- Level III: Students will understand the concepts and be able to work independently.

All children learn at their own rate. Use your own judgment for introducing concepts to children when developmentally appropriate.

Hands-On Learning

Review is an important part of learning. It helps to ensure that skills are not only covered but are internalized. The flash cards at the back of this book will offer endless opportunities for review. Use them for a basic math facts drill, or to play bingo or other fun games.

There is also a certificate template at the back of this book for use as students excel at daily assignments or when they finish a unit.

Common Core State Standards Alignment Chart

Common Core State Standards*		Practice Page(s)
Operations and Algebraic Thinking		
Write and interpret numerical expressions.	5.OA.1–5.OA.2	5–10
Analyze patterns and relationships.	5.OA.3	98–100
Number and Operations in Base Ten		
Understand the place value system.	5.NBT.1–5.NBT.4	11–19
Perform operations with multi-digit whole numbers and with decimals to hundredths.	5.NBT.5–5.NBT.7	20–40, 53–55
Number and Operations—Fractions		
Use equivalent fractions as a strategy to add and subtract fractions.	5.NF.1–5.NF.2	41–52, 71
Apply and extend previous understandings of multiplication and division to multiply and divide fractions	5.NF.3–5.NF.7	53–70, 72, 73
Measurement and Data		
Convert like measurement units within a given measurement system.	5.MD.1	74–76
Represent and interpret data.	5.MD.2	77–79
Geometric measurement: understand concepts of volume and relate volume to multiplication and to addition.	5.MD.3–5.MD.5	80–94
Geometry		
Graph points on the coordinate plane to solve real-world and mathematical problems.	5.G.1–5.G.2	95–100
Classify two-dimensional figures into categories based on their properties.	5.G.3–5.G.4	101–103

Evaluating Numerical Expressions

Use the following order to calculate and solve expressions:

1.   Solve inside parentheses. $(36 \div 12) \times 2 + 3 \longrightarrow (3) \times 2 + 3$

2.   Multiply and divide from left to right. $3 \times 2 + 3 \longrightarrow 6 + 3$

3.   Add and subtract from left to right. $6 + 3 \longrightarrow 9$

Solve each expression. Remember to follow the order of operations.

1. $(6 \times 2) + 8 =$ _____

2. $3 + (8 \times 2) =$ _____

3. $14 \div 2 + 3 =$ _____

4. $21 \div 7 \times 2 =$ _____

5. $(5 \times 2) + 3 =$ _____

6. $(10 + 10) \div 2 =$ _____

7. $6 \times (3 + 3) =$ _____

8. $10 \times 10 \div 25 =$ _____

9. $(17 - 7) \div 5 =$ _____

10. $50 \div 5 + 3 =$ _____

Evaluating Numerical Expressions

Solve each expression. Remember to follow the order of operations.

1. $8 - 2 + 9 = $ _____

2. $18 \div 9 + 3 = $ _____

3. $(15 \div 3) \times 5 = $ _____

4. $(45 \div 9) \times 4 = $ _____

5. $5 + (21 + 4) = $ _____

6. $15 - 2 + 8 = $ _____

7. $40 \div 5 - 7 = $ _____

8. $(5 \times 7) - 30 = $ _____

9. $(4 \times 4) + 5 = $ _____

10. $21 \div 7 + 3 = $ _____

11. $(2^2 - 2) \times 3 + 8 = $ _____

12. $(15 \times 2) \div 10 + 8 = $ _____

13. $10 - (14 \div 2) + 3 = $ _____

14. $2 \times 2 \div 2 \times 8 = $ _____

Evaluating Numerical Expressions

Solve each expression.

1. $8 \times (8 - 2) + 6^2 = $ _____

2. $(52 - 4^2) \div (8 - 4) = $ _____

3. $(2 + 2)^2 + (14 \div 2) = $ _____

4. $(12 \times 10 - 4^2) - 5 = $ _____

5. $(14 - 2)^2 + (8 \div 4) = $ _____

6. $(64 - 6^2) \div (9 - 2) = $ _____

7. $6 \times (13 - 4) + 2^2 = $ _____

8. $[4^2 + (15 \div 5 + 5^2)] + 4^2 = $ _____

9. $[5^2 + (10 \div 2 + 2^2)] - 5^2 = $ _____

10. $[(9 - 2)^2 + 3] + 5 + 3^2 = $ _____

11. $[(10 - 5)^2 \times 7] + 5 + 5^2 = $ _____

12. $6 + [9 + (9 - 3)^2] + 7 = $ _____

13. $(16 \div 8)^2 + [(15 - 2) \times 3^2] = $ _____

14. $13 + [7 + (10 - 6)^2] - 8 = $ _____

Interpreting Numerical Expressions

Expressions can be written with numbers and symbols or in words.

4 more than the product of 6 and 7   add 10 and 12, then divide in half

$$4 + (6 \times 7)$$ $$(10 + 12) \div 2$$

Look for key words to help you decide which operations to use. Use parentheses to group the part of the expression that should happen first.

Write each expression with numbers.

1. 3 times the sum of 2 and 46 _____

2. 16 more than the product of 2 and 9 _____

3. subtract 4 from 29, then double _____

4. 6 less than the quotient of 90 divided by 9 _____

Write each expression in words.

5. $9 + (24 \div 6)$ _____

6. $(86 - 72) + 6$ _____

7. $(22 \times 3) \div 2$ _____

8. $4 \times (5 + 83)$ _____

Interpreting Numerical Expressions

Write each expression with numbers.

1. add 34 to itself, then divide in half _____

2. subtract 6 from 34, then triple _____

3. 5 times the sum of 23 and 7 _____

4. the quotient of 100 and double 25 _____

5. 8 more than the product of 12 and 4 _____

6. 16 less than the quotient of 200 divided by 10 _____

Write each expression in words.

7. $(33 + 6) \div 3$ _____

8. $2 \times (15 + 67)$ _____

9. $(3 \times 13) + 2$ _____

10. $(20 - 4) + 100$ _____

11. $(100 - 45) \times 2$ _____

12. $70 + (45 \div 9)$ _____

Interpreting Numerical Expressions

Write each expression with numbers.

1. half of the product of 42 and 3 _____

2. four times the sum of 53 and 11 _____

3. the product of 7 and the sum of 5 and 74 _____

4. 23 times 10, divided by 5 _____

5. 18 more than the product of 2 and 42 _____

6. 14 less than the product of 16 and 10 _____

7. the sum of 45 and the product of 3 and 12 _____

Write each expression in words.

8. $(13 \times 6) \div 2$ _____

9. $12 + (30 + 7)$ _____

10. $(40 \div 8) - 12$ _____

11. $(21 + 9) \div 3$ _____

12. $(10 + 2) - (45 \div 9)$ _____

_____

13. $80 - (4 \times 4)$ _____

_____

Name _____

Powers of Ten

Numbers can be abbreviated using exponential notation.

An exponent tells how many times a factor is multiplied by itself.

$10^3 = 10 \times 10 \times 10 = 1,000$ 10 is multiplied by itself 3 times.

Look for patterns when a power of 10 is multiplied by another number between 1 and 9.

$7,000,000 = 7 \times 10^6$

Hint: To know what power of 10 to use, simply match the power of 10 to the number of zeros in the number.

$4,\underline{000} = 4 \times 10^3$ $9\underline{00,000} = 9 \times 10^5$

Write each number with an exponent.

1. 10 to the fourth power _____ 2. 10 to the third power _____ 3. 10 to the eighth power _____

Solve.

4. $10^3 =$ _____ $10^6 =$ _____ $10^4 =$ _____

5. $10^2 =$ _____ $10^{10} =$ _____ $10^8 =$ _____

6. $10^7 =$ _____ $10^5 =$ _____ $10^9 =$ _____

Rewrite each problem without the exponent. Then, solve.

7. $3 \times 10^2 =$ _____ = _____ $8 \times 10^3 =$ _____ = _____

8. $6 \times 10^4 =$ _____ = _____ $4 \times 10^5 =$ _____ = _____

Write each number as a number multiplied by a power of 10.

9. 7,000 = _____ 5,000 = _____ 600,000 = _____

10. 8,000,000 = _____ 40,000 = _____ 3,000,000,000 = _____

Powers of Ten

Choose numbers to complete each problem. Solve. Look for patterns.

1. _____ × 10 = _____   2. _____ × 100 = _____   3. _____ × 1,000 = _____

   _____ × 10 = _____   _____ × 100 = _____   _____ × 1,000 = _____

   _____ × 10 = _____   _____ × 100 = _____   _____ × 1,000 = _____

   _____ ÷ 10 = _____   _____ ÷ 100 = _____   _____ ÷ 1,000 = _____

   _____ ÷ 10 = _____   _____ ÷ 100 = _____   _____ ÷ 1,000 = _____

   _____ ÷ 10 = _____   _____ ÷ 100 = _____   _____ ÷ 1,000 = _____

4. _____ × 0.1 = _____   5. _____ × 0.01 = _____   6. _____ × 0.001 = _____

   _____ × 0.1 = _____   _____ × 0.01 = _____   _____ × 0.001 = _____

   _____ × 0.1 = _____   _____ × 0.01 = _____   _____ × 0.001 = _____

   _____ ÷ 0.1 = _____   _____ ÷ 0.01 = _____   _____ ÷ 0.001 = _____

   _____ ÷ 0.1 = _____   _____ ÷ 0.01 = _____   _____ ÷ 0.001 = _____

   _____ ÷ 0.1 = _____   _____ ÷ 0.01 = _____   _____ ÷ 0.001 = _____

7. Write a rule for multiplying and dividing by powers of 10.

   _____

   _____

   _____

   _____

Powers of Ten

Use what you know about multiplying and dividing by powers of 10 to answer each problem without multiplying or dividing.

1. $6 \times 10,000 =$ _____

   $6 \times 1,000 =$ _____

   $6 \times 100 =$ _____

   $6 \times 10 =$ _____

   $6 \times 1 =$ _____

   $6 \div 1 =$ _____

   $6 \div 10 =$ _____

   $6 \div 100 =$ _____

   $6 \div 1,000 =$ _____

2. $8 \times 10,000 =$ _____

   $8 \times 1,000 =$ _____

   $8 \times 100 =$ _____

   $8 \times 10 =$ _____

   $8 \times 1 =$ _____

   $8 \div 1 =$ _____

   $8 \div 10 =$ _____

   $8 \div 100 =$ _____

   $8 \div 1,000 =$ _____

3. $24 \times 10,000 =$ _____

   $24 \times 1,000 =$ _____

   $24 \times 100 =$ _____

   $24 \times 10 =$ _____

   $24 \times 1 =$ _____

   $24 \div 1 =$ _____

   $24 \div 10 =$ _____

   $24 \div 100 =$ _____

   $24 \div 1,000 =$ _____

4. $13 \times 10,000 =$ _____

   $13 \times 1,000 =$ _____

   $13 \times 100 =$ _____

   $13 \times 10 =$ _____

   $13 \times 1 =$ _____

   $13 \div 1 =$ _____

   $13 \div 10 =$ _____

   $13 \div 100 =$ _____

   $13 \div 1,000 =$ _____

Name _____

Understanding Decimals

$2\frac{4}{10}$

What portion of these boxes are shaded? two entire boxes

What portion of this box is shaded? four-tenths of the box

2.4 (two and four-tenths)

This can be spoken, "two point four," or "two and four-tenths."

Note: When writing a decimal, if there are no whole numbers, place a zero left of the decimal point. Examples: seven-tenths = 0.7, nine-tenths = 0.9

Write each decimal.

1. three and five-tenths _____

2. six and one-tenth _____

3. eight-tenths _____

4. eight and three-tenths _____

5. three-tenths _____

6. two and one-tenth _____

7. seven-tenths _____

8. twenty and two-tenths _____

9. four-tenths _____

10. thirty-seven and two-tenths _____

Write each decimal in words.

11. 3.9 _____

12. 2.7 _____

13. 12.8 _____

14. 7.3 _____

Use <, >, or = to compare the decimals.

15. 3.4 ◯ 4.5   16. 6.01 ◯ 2.06   17. 5.01 ◯ 51.09   18. 3.02 ◯ 2.03

Understanding Decimals

Write each decimal.

1. one-tenth _____

2. twenty-seven hundredths _____

3. three-thousandths _____

4. seven-tenths _____

5. forty-five hundredths _____

6. fifty-one thousandths _____

7. four hundred and one-tenth _____

8. fifty-five and three-tenths _____

9. six-tenths _____

10. one-hundredth _____

Write each decimal in words.

11. 0.04 _____

12. 0.99 _____

13. 0.8 _____

14. 4.89 _____

15. 0.06 _____

Order the numbers in each series from least to greatest.

16. 1.87, 0.187, 10.87

17. 0.045, 0.45, 0.04

18. 0.0065, 0.06, 0.006

19. 0.91, 0.44, 0.23

20. 6.07, 6.17, 6.37

21. 8.98, 8.89, 8.9

Understanding Decimals

Write each decimal.

1. three hundred and
    seven-hundredths _____

2. fifteen and
    forty-five thousandths _____

3. two hundred eighteen
    and four-thousandths _____

4. two-thousandths _____

5. sixty-seven and six hundred
    thirty-one thousandths _____

6. twelve-thousandths _____

7. forty-nine and
    ninety-nine thousandths _____

8. five and eight hundred
    forty-five thousandths _____

9. eight-thousandths _____

10. ten and six hundred
    two-thousandths _____

Write each decimal in words.

11. 0.035 _____

12. 89.004 _____

13. 324.008 _____

14. 72.045 _____

Use <, >, or = to compare the decimals.

15. 0.567 ◯ 0.423   16. 56.001 ◯ 56.01   17. 3.003 ◯ 33.003   18. 5.9 ◯ 5.09

19. 0.987 ◯ 0.789   20. 1.456 ◯ 1.665   21. 2.076 ◯ 8.076   22. 2.798 ◯ 3.009

Rounding Decimals

To round a decimal, follow these steps:

1. Underline the place value you are rounding to.

2. If the number to the right of the underline is 0, 1, 2, 3, or 4, the underlined digit stays the same. All of the digits to the right change to zeros.

3. If the number to the right of the underline is 5, 6, 7, 8, or 9, the underlined digit goes up by one. All of the digits to the right change to zeros.

Examples:
    Round to the nearest whole number: 4.8 rounds up to 5.0
    Round to the nearest tenth: 14.24 rounds down to 14.20

Round to the nearest whole number.

1. 3.67 _____
2. 6.8 _____
3. 11.4 _____
4. 5.9 _____

5. 21.24 _____
6. 10.51 _____
7. 4.9 _____
8. 14.2 _____

9. 8.6 _____
10. 7.8 _____
11. 9.21 _____
12. 10.9 _____

13. 9.7 _____
14. 10.3 _____
15. 8.3 _____
16. 7.4 _____

Round to the nearest tenth.

17. 6.29 _____
18. 10.68 _____
19. 14.83 _____
20. 6.84 _____

21. 3.48 _____
22. 24.37 _____
23. 17.47 _____
24. 28.15 _____

25. 5.49 _____
26. 10.43 _____
27. 3.56 _____
28. 6.26 _____

29. 17.64 _____
30. 112.26 _____
31. 9.42 _____
32. 400.67 _____

Rounding Decimals

> Remember, to round a decimal, underline the place value you are rounding to. If the digit to the right is 4 or less, round down. If the digit is 5 or greater, round up.

Round to the nearest whole number.

1. 15.4 _____
2. 3.9 _____
3. 58.7 _____
4. 294.2 _____

5. 27.3 _____
6. 18.8 _____
7. 93.4 _____
8. 914.5 _____

Round to the nearest tenth.

9. 25.25 _____
10. 8.16 _____
11. 93.12 _____
12. 724.95 _____

13. 289.37 _____
14. 20.83 _____
15. 204.47 _____
16. 38.52 _____

Round to the nearest hundredth.

17. 205.602 _____
18. 9.995 _____
19. 39.194 _____
20. 532.364 _____

21. 34.436 _____
22. 4.783 _____
23. 93.295 _____
24. 59.364 _____

Rounding Decimals

Round to the nearest whole number.

1. 45.678 _____
2. 2.68 _____
3. 612.123 _____

4. 2,345.5 _____
5. 7.29 _____
6. 1.398 _____

7. 87.056 _____
8. 43.089 _____
9. 234.768 _____

10. 567.289 _____
11. 7.8 _____
12. 6.29 _____

13. 89.519 _____
14. 4.916 _____
15. 0.897 _____

Round to the nearest tenth.

16. 4.379 _____
17. 2.819 _____
18. 543.18 _____

19. 56.14 _____
20. 3.157 _____
21. 78.028 _____

22. 0.417 _____
23. 36.192 _____
24. 1.248 _____

25. 34.248 _____
26. 8.77 _____
27. 16.651 _____

28. 0.54 _____
29. 678.456 _____
30. 0.25 _____

Round to the nearest hundredth.

31. 34.248 _____
32. 5.251 _____
33. 6.108 _____

34. 5.213 _____
35. 9.178 _____
36. 23.682 _____

37. 2.454 _____
38. 9.017 _____
39. 7.271 _____

40. 6.319 _____
41. 2,345.124 _____
42. 6.237 _____

43. 45.814 _____
44. 38.199 _____
45. 435.458 _____

Multiplying Multi-Digit Numbers

Solve each problem. Regroup when necessary.

1. $\begin{array}{r} 323 \\ \times5 \\ \hline \end{array}$
2. $\begin{array}{r} 515 \\ \times4 \\ \hline \end{array}$
3. $\begin{array}{r} 255 \\ \times4 \\ \hline \end{array}$
4. $\begin{array}{r} 915 \\ \times2 \\ \hline \end{array}$
5. $\begin{array}{r} 860 \\ \times2 \\ \hline \end{array}$
6. $\begin{array}{r} 561 \\ \times9 \\ \hline \end{array}$

7. $\begin{array}{r} 109 \\ \times4 \\ \hline \end{array}$
8. $\begin{array}{r} 812 \\ \times8 \\ \hline \end{array}$
9. $\begin{array}{r} 503 \\ \times3 \\ \hline \end{array}$
10. $\begin{array}{r} 827 \\ \times3 \\ \hline \end{array}$
11. $\begin{array}{r} 122 \\ \times8 \\ \hline \end{array}$
12. $\begin{array}{r} 523 \\ \times6 \\ \hline \end{array}$

13. $\begin{array}{r} 5,306 \\ \times3 \\ \hline \end{array}$
14. $\begin{array}{r} 6,241 \\ \times7 \\ \hline \end{array}$
15. $\begin{array}{r} 6,384 \\ \times9 \\ \hline \end{array}$
16. $\begin{array}{r} 4,634 \\ \times2 \\ \hline \end{array}$
17. $\begin{array}{r} 8,436 \\ \times5 \\ \hline \end{array}$
18. $\begin{array}{r} 5,691 \\ \times5 \\ \hline \end{array}$

19. $\begin{array}{r} 35 \\ \times28 \\ \hline \end{array}$
20. $\begin{array}{r} 73 \\ \times56 \\ \hline \end{array}$
21. $\begin{array}{r} 72 \\ \times43 \\ \hline \end{array}$
22. $\begin{array}{r} 63 \\ \times58 \\ \hline \end{array}$
23. $\begin{array}{r} 83 \\ \times27 \\ \hline \end{array}$
24. $\begin{array}{r} 70 \\ \times60 \\ \hline \end{array}$

Multiplying Multi-Digit Numbers

Solve each problem. Regroup when necessary.

1. $\begin{array}{r} 87 \\ \times\ \ 5 \\ \hline \end{array}$
2. $\begin{array}{r} 72 \\ \times\ 18 \\ \hline \end{array}$
3. $\begin{array}{r} 425 \\ \times\ \ 15 \\ \hline \end{array}$
4. $\begin{array}{r} 303 \\ \times\ \ 83 \\ \hline \end{array}$
5. $\begin{array}{r} 187 \\ \times\ \ 26 \\ \hline \end{array}$
6. $\begin{array}{r} 93 \\ \times\ \ 6 \\ \hline \end{array}$

7. $\begin{array}{r} 63 \\ \times\ 25 \\ \hline \end{array}$
8. $\begin{array}{r} 313 \\ \times\ \ 72 \\ \hline \end{array}$
9. $\begin{array}{r} 442 \\ \times\ \ 81 \\ \hline \end{array}$
10. $\begin{array}{r} 593 \\ \times\ \ 45 \\ \hline \end{array}$
11. $\begin{array}{r} 84 \\ \times\ \ 3 \\ \hline \end{array}$
12. $\begin{array}{r} 42 \\ \times\ \ 8 \\ \hline \end{array}$

13. $\begin{array}{r} 81 \\ \times\ 53 \\ \hline \end{array}$
14. $\begin{array}{r} 872 \\ \times\ \ 20 \\ \hline \end{array}$
15. $\begin{array}{r} 351 \\ \times\ \ 67 \\ \hline \end{array}$
16. $\begin{array}{r} 52 \\ \times\ \ 4 \\ \hline \end{array}$
17. $\begin{array}{r} 75 \\ \times\ 21 \\ \hline \end{array}$
18. $\begin{array}{r} 21 \\ \times\ 10 \\ \hline \end{array}$

19. $\begin{array}{r} 214 \\ \times\ \ 87 \\ \hline \end{array}$
20. $\begin{array}{r} 109 \\ \times\ \ 15 \\ \hline \end{array}$
21. $\begin{array}{r} 12 \\ \times\ \ 9 \\ \hline \end{array}$
22. $\begin{array}{r} 16 \\ \times\ \ 8 \\ \hline \end{array}$
23. $\begin{array}{r} 87 \\ \times\ 26 \\ \hline \end{array}$
24. $\begin{array}{r} 99 \\ \times\ 21 \\ \hline \end{array}$

Multiplying Multi-Digit Numbers

Solve each problem.

1. 918
    × 55

2. 755
    × 221

3. 618
    × 500

4. 1,242
    × 687

5. 622
    × 437

6. 832
    × 106

7. 391
    × 125

8. 3,861
    × 392

9. 518
    × 42

10. 535
    × 391

11. 663
    × 482

12. 4,369
    × 873

13. 925
    × 54

14. 851
    × 462

15. 528
    × 331

16. 7,421
    × 694

17. 622
    × 33

18. 795
    × 787

19. 683
    × 435

20. 5,872
    × 515

Division with One-Digit Divisors

Solve each problem.

1. $4\overline{)100}$ 2. $2\overline{)132}$ 3. $3\overline{)225}$ 4. $9\overline{)198}$

5. $2\overline{)902}$ 6. $7\overline{)112}$ 7. $6\overline{)510}$ 8. $4\overline{)216}$

9. $6\overline{)426}$ 10. $2\overline{)630}$ 11. $3\overline{)138}$ 12. $9\overline{)369}$

13. $8\overline{)624}$ 14. $6\overline{)396}$ 15. $8\overline{)648}$ 16. $5\overline{)310}$

17. $5\overline{)425}$ 18. $7\overline{)672}$ 19. $3\overline{)864}$ 20. $7\overline{)966}$

Division with One-Digit Divisors

Solve each problem.

1. $9\overline{)1,368}$ 2. $4\overline{)1,228}$ 3. $8\overline{)5,392}$ 4. $6\overline{)1,878}$

5. $5\overline{)1,395}$ 6. $7\overline{)2,926}$ 7. $4\overline{)1,008}$ 8. $5\overline{)975}$

9. $4\overline{)2,128}$ 10. $2\overline{)1,224}$ 11. $6\overline{)2,706}$ 12. $3\overline{)2,019}$

13. $3\overline{)1,008}$ 14. $8\overline{)3,888}$ 15. $7\overline{)1,421}$ 16. $5\overline{)1,125}$

17. $2\overline{)1,024}$ 18. $3\overline{)1,134}$ 19. $8\overline{)4,960}$ 20. $9\overline{)2,790}$

Division with One-Digit Divisors

Solve each problem.

1. 4)873

2. 5)943

3. 8)957

4. 9)987

5. 7)915

6. 5)527

7. 2)597

8. 9)973

9. 4)574

10. 6)653

11. 3)784

12. 4)486

13. 3)629

14. 2)301

15. 5)637

16. 4)862

17. 2)733

18. 8)937

19. 3)574

20. 4)653

Division with Two-Digit Divisors

Solve each problem.

1. $32\overline{)512}$ 2. $52\overline{)624}$ 3. $18\overline{)450}$ 4. $32\overline{)768}$

5. $62\overline{)992}$ 6. $41\overline{)820}$ 7. $12\overline{)144}$ 8. $32\overline{)960}$

9. $18\overline{)702}$ 10. $39\overline{)858}$ 11. $15\overline{)540}$ 12. $23\overline{)345}$

13. $56\overline{)952}$ 14. $47\overline{)517}$ 15. $27\overline{)810}$ 16. $26\overline{)338}$

17. $25\overline{)350}$ 18. $45\overline{)990}$ 19. $24\overline{)600}$ 20. $54\overline{)864}$

Division with Two-Digit Divisors

Solve each problem.

1. $67\overline{)807}$ 2. $37\overline{)369}$ 3. $64\overline{)654}$ 4. $81\overline{)921}$

5. $61\overline{)741}$ 6. $58\overline{)368}$ 7. $18\overline{)652}$ 8. $23\overline{)875}$

9. $13\overline{)235}$ 10. $40\overline{)142}$ 11. $25\overline{)465}$ 12. $11\overline{)505}$

13. $19\overline{)410}$ 14. $32\overline{)458}$ 15. $53\overline{)367}$ 16. $45\overline{)787}$

17. $22\overline{)268}$ 18. $42\overline{)632}$ 19. $56\overline{)647}$ 20. $87\overline{)357}$

Division with Two-Digit Divisors

Solve each problem.

1. $43 \overline{)1,256}$ 2. $48 \overline{)2,541}$ 3. $65 \overline{)1,596}$ 4. $22 \overline{)7,321}$

5. $21 \overline{)3,010}$ 6. $39 \overline{)8,563}$ 7. $82 \overline{)4,512}$ 8. $37 \overline{)2,148}$

9. $30 \overline{)6,172}$ 10. $78 \overline{)5,000}$ 11. $77 \overline{)2,159}$ 12. $85 \overline{)3,578}$

13. $59 \overline{)8,787}$ 14. $55 \overline{)9,999}$ 15. $27 \overline{)3,265}$ 16. $56 \overline{)5,892}$

Adding Decimals

Solve each problem. Regroup when necessary.

1. 14.2   2. 18.7   3. 6.54   4. 15.2   5. 16.6   6. 9.41
   + 12.1   + 10.5   + 1.47   + 12.3   + 13.8   + 7.85

7. 18.2   8. 15.2   9. 3.94   10. 22.2   11. 14.9   12. 7.54
   + 16.5   + 13.0   + 2.22   + 13.1   + 12.0   + 2.24

13. 47.5   14. 49.4   15. 8.85   16. 54.8   17. 4.58
    + 32.6   + 11.1   + 7.33   + 13.2   + 2.31

18. 12.95 + 5.06 =   19. 13.8 + 6.9 =

20. 46.02 + 75.67 =   21. 16.3 + 35.7 =

22. 3.25 + 3.25 =   23. 87.01 + 16.53 =

Name _____

Adding Decimals

Solve each problem. Regroup when necessary.

1. 2.4
  + 1.7

2. 18.6
  + 9.5

3. 0.01
  + 0.72

4. 3.2
    1.4
  + 7.8

5. 2.01
    3.09
  + 8.62

6. 8.1
  + 9.2

7. 14.3
  + 1.9

8. 1.04
  + 2.07

9. 86.7
    5.2
  + 8.4

10. 42.65
     67.23
  + 12.12

11. 10.3
  + 7.4

12. 24.7
  + 32.6

13. 16.52
  + 13.63

14. 9.1
    12.5
  + 19.4

15. 492.6
    382.3
  + 225.7

16. 1.5
  + 1.5

17. 20.5
  + 32.3

18. 14.87
  + 56.09

19. 40.08
    60.27
  + 50.33

20. 4.08
    1.38
  + 0.06

30

© Carson-Dellosa • CD-104630

Adding Decimals

Solve each problem.

1.
```
   2.34
   0.02
+ 1.65
```

2.
```
  543.7
    3.42
+   0.06
```

3.
```
  72.56
  12.38
+  0.07
```

4.
```
  22.87
  45.7
+  1.26
```

5.
```
  987.5
    4.1
+  30.2
```

6.
```
   2.14
   0.07
+ 72.4
```

7.
```
    1.70
   23.75
+   0.05
```

8.
```
  86.15
   0.07
+  5.72
```

9.
```
   5.1
   7.53
+ 87.4
```

10.
```
  0.2
  1.2
+ 0.12
```

11.
```
   1.45
  20.03
+ 0.17
```

12.
```
   4.5
   5.4
+ 12.67
```

13.
```
  42.7
   0.03
+  1.7
```

14.
```
   87.5
    1.2
+ 591.35
```

15.
```
  0.72
  1.33
+ 12
```

16.
```
  42
   0.54
+  7.8
```

Subtracting Decimals

Solve each problem. Regroup when necessary.

1. $\begin{array}{r} 5.6 \\ -\ 3.2 \\ \hline \end{array}$

2. $\begin{array}{r} 10.4 \\ -\ 8.2 \\ \hline \end{array}$

3. $\begin{array}{r} 8.5 \\ -\ 3.5 \\ \hline \end{array}$

4. $\begin{array}{r} 7.8 \\ -\ 4.5 \\ \hline \end{array}$

5. $\begin{array}{r} 9.3 \\ -\ 7.5 \\ \hline \end{array}$

6. $\begin{array}{r} 86.5 \\ -\ 2.3 \\ \hline \end{array}$

7. $\begin{array}{r} 6.3 \\ -\ 4.1 \\ \hline \end{array}$

8. $\begin{array}{r} 8.7 \\ -\ 5.2 \\ \hline \end{array}$

9. $\begin{array}{r} 9.65 \\ -\ 4.22 \\ \hline \end{array}$

10. $\begin{array}{r} 8.6 \\ -\ 5.2 \\ \hline \end{array}$

11. $\begin{array}{r} 16.4 \\ -\ 8.2 \\ \hline \end{array}$

12. $\begin{array}{r} 75.4 \\ -\ 3.1 \\ \hline \end{array}$

13. $\begin{array}{r} 7.6 \\ -\ 3.2 \\ \hline \end{array}$

14. $\begin{array}{r} 26.7 \\ -\ 2.5 \\ \hline \end{array}$

15. $\begin{array}{r} 16.2 \\ -\ 4.1 \\ \hline \end{array}$

16. $72.5 - 63.7 =$

17. $8.1 - 6.5 =$

Subtracting Decimals

Solve each problem. Regroup when necessary.

1. $\begin{array}{r} 326.7 \\ -\ \ 42.8 \\ \hline \end{array}$

2. $\begin{array}{r} 14.021 \\ -\ \ \ \ 5.6 \\ \hline \end{array}$

3. $\begin{array}{r} 1.58 \\ -\ 0.75 \\ \hline \end{array}$

4. $\begin{array}{r} 16.88 \\ -\ \ 9.3 \\ \hline \end{array}$

5. $\begin{array}{r} 52.07 \\ -\ \ 3.9 \\ \hline \end{array}$

6. $\begin{array}{r} 7.57 \\ -\ 6.85 \\ \hline \end{array}$

7. $\begin{array}{r} 8.12 \\ -\ 6.01 \\ \hline \end{array}$

8. $\begin{array}{r} 18.9 \\ -\ 16.42 \\ \hline \end{array}$

9. $\begin{array}{r} 1.97 \\ -\ 1.68 \\ \hline \end{array}$

10. $\begin{array}{r} 14.9 \\ -\ 3.2 \\ \hline \end{array}$

11. $19.5 - 0.01 =$

12. $0.51 - 0.32 =$

13. $42.62 - 10.35 =$

14. $28.4 - 4.62 =$

15. $33.45 - 15.4 =$

16. $18.5 - 9.5 =$

Subtracting Decimals

Solve each problem.

1.   19.86
   −   1.07

2.   4.52
   − 0.4

3.   6.25
   − 3.01

4.   23.15
   −   3.08

5.   0.7
   − 0.506

6.   20.34
   −   0.3

7.   756.83
   −   22.5

8.   38.7
   − 5.21

9.   1.42
   − 1.2

10.   71.34
   −   2.67

11.   31.1
   − 3.05

12.   0.65
   − 0.224

13.   2.3
   − 1.43

14.   32.45
   −   1.2

15.   81.38
   −   2.77

16.   24.75
   −   6.24

Multiplying Decimals

Solve each problem. Regroup when necessary.

1. $\begin{array}{r} 5.2 \\ \times\ 1.8 \\ \hline \end{array}$

2. $\begin{array}{r} 10.5 \\ \times\ 6.6 \\ \hline \end{array}$

3. $\begin{array}{r} 2.8 \\ \times\ 9.9 \\ \hline \end{array}$

4. $\begin{array}{r} 2.2 \\ \times\ 4.4 \\ \hline \end{array}$

5. $\begin{array}{r} 0.12 \\ \times\ 3.7 \\ \hline \end{array}$

6. $\begin{array}{r} 5.2 \\ \times\ 0.2 \\ \hline \end{array}$

7. $\begin{array}{r} 1.3 \\ \times\ 1.0 \\ \hline \end{array}$

8. $\begin{array}{r} 7.1 \\ \times\ 0.25 \\ \hline \end{array}$

9. $\begin{array}{r} 7.5 \\ \times\ 2.7 \\ \hline \end{array}$

10. $\begin{array}{r} 6.4 \\ \times\ 2.5 \\ \hline \end{array}$

11. $\begin{array}{r} 16.2 \\ \times\ 1.1 \\ \hline \end{array}$

12. $\begin{array}{r} 2.0 \\ \times\ 2.1 \\ \hline \end{array}$

13. $\begin{array}{r} 5.4 \\ \times\ 1.3 \\ \hline \end{array}$

14. $\begin{array}{r} 6.6 \\ \times\ 1.5 \\ \hline \end{array}$

15. $\begin{array}{r} 0.44 \\ \times\ 0.1 \\ \hline \end{array}$

16. $\begin{array}{r} 0.34 \\ \times\ 0.12 \\ \hline \end{array}$

17. $\begin{array}{r} 5.5 \\ \times\ 4.6 \\ \hline \end{array}$

18. $\begin{array}{r} 6.1 \\ \times\ 2.5 \\ \hline \end{array}$

19. $\begin{array}{r} 5.6 \\ \times\ 7.3 \\ \hline \end{array}$

20. $\begin{array}{r} 3.3 \\ \times\ 0.8 \\ \hline \end{array}$

Multiplying Decimals

Solve each problem. Round to the nearest thousandth when necessary.

1.   0.2
   × 4

2.   0.08
   × 6

3.   5.7
   × 0.9

4.   0.14
   × 0.27

5.   0.67
   × 5.4

6.   0.8
   × 3

7.   0.04
   × 5

8.   4.3
   × 2

9.   7.2
   × 5.3

10.   7.1
   × 5.5

11.   0.5
   × 0.4

12.   6
   × 0.12

13.   1.07
   × 0.55

14.   0.09
   × 0.06

15.   0.07
   × 0.15

16.   0.23
   × 0.7

17.   0.5
   × 0.25

18.   2.15
   × 0.8

19.   1.3
   × 3.1

20.   32
   × 6.4

Multiplying Decimals

Solve each problem. Round to the nearest thousandth when necessary.

1.
$$0.18 \times 1.5$$

2.
$$0.16 \times 100$$

3.
$$0.08 \times 0.42$$

4.
$$87.85 \times 63.4$$

5.
$$536.7 \times 1.79$$

6.
$$6.42 \times 3.7$$

7.
$$0.48 \times 13.5$$

8.
$$0.65 \times 53.7$$

9.
$$4.06 \times 0.7$$

10.
$$43.6 \times 64.7$$

11.
$$80.42 \times 7.86$$

12.
$$0.62 \times 5.97$$

13.
$$84.4 \times 0.07$$

14.
$$5.11 \times 0.78$$

15.
$$4.35 \times 0.68$$

16.
$$3.17 \times 0.78$$

Dividing Decimals

Solve each problem.

1. $2\overline{)8.44}$

2. $5\overline{)1.25}$

3. $14\overline{)7.21}$

4. $2\overline{)38.6}$

5. $67\overline{)656.6}$

6. $52\overline{)166.4}$

7. $7\overline{)3.92}$

8. $2\overline{)9.4}$

9. $24\overline{)17.28}$

10. $6\overline{)3.6}$

11. $7\overline{)3.92}$

12. $46\overline{)234.6}$

13. $3\overline{)42.3}$

14. $5\overline{)72.55}$

15. $10\overline{)16.6}$

16. $4\overline{)9.6}$

17. $5\overline{)8.65}$

18. $67\overline{)274.7}$

Dividing Decimals

Solve each problem.

1. $9\overline{)2.7}$

2. $7\overline{)2.1}$

3. $4\overline{)0.16}$

4. $0.8\overline{)56}$

5. $6\overline{)3.6}$

6. $8\overline{)0.64}$

7. $9\overline{)0.27}$

8. $0.07\overline{)2.1}$

9. $3\overline{)2.7}$

10. $6\overline{)0.30}$

11. $0.04\overline{)28}$

12. $0.9\overline{)5.4}$

13. $3\overline{)0.18}$

14. $2\overline{)0.12}$

15. $0.9\overline{)72}$

16. $0.7\overline{)0.35}$

17. $4\overline{)2.4}$

18. $5\overline{)2.5}$

19. $0.04\overline{)36}$

20. $0.9\overline{)6.3}$

Dividing Decimals

Solve each problem.

1. $0.8\overline{)64}$

2. $0.25\overline{)100}$

3. $6.1\overline{)7.93}$

4. $0.5\overline{)35}$

5. $1.2\overline{)48}$

6. $5.3\overline{)42.4}$

7. $0.4\overline{)64}$

8. $0.7\overline{)4.9}$

9. $0.19\overline{)15.2}$

10. $0.3\overline{)9}$

11. $9.6\overline{)82.8}$

12. $0.17\overline{)3.23}$

13. $0.8\overline{)152}$

14. $0.3\overline{)0.63}$

15. $2.1\overline{)13.65}$

16. $0.12\overline{)360}$

17. $0.23\overline{)21.85}$

18. $7.2\overline{)40.32}$

Adding Fractions with Unlike Denominators

<table>
<tr><td>1. Find the least common denominator (LCD).

$\frac{1}{3} + \frac{1}{4}$

LCD = 12</td><td>2. Find the equivalent fractions.

$\frac{1}{3} \times \frac{4}{4} = \frac{4}{12}$

$\frac{1}{4} \times \frac{3}{3} = \frac{3}{12}$</td><td>3. Add the numerators.

$\frac{4}{12}$
$+ \frac{3}{12}$
―――
$\frac{7}{12}$</td></tr>
</table>

Solve each problem. Write each answer in simplest form.

1. $\frac{1}{4}$
    $+ \frac{3}{5}$

2. $\frac{1}{2}$
    $+ \frac{1}{5}$

3. $\frac{4}{5}$
    $+ \frac{2}{3}$

4. $\frac{2}{3}$
    $+ \frac{2}{5}$

5. $\frac{4}{5}$
    $+ \frac{7}{8}$

6. $\frac{3}{4}$
    $+ \frac{1}{3}$

7. $\frac{1}{6}$
    $+ \frac{2}{5}$

8. $\frac{3}{6}$
    $+ \frac{3}{4}$

9. $\frac{1}{4}$
    $+ \frac{2}{3}$

10. $\frac{5}{8}$
     $+ \frac{2}{3}$

11. $\frac{2}{5}$
     $+ \frac{1}{3}$

12. $\frac{1}{3}$
     $+ \frac{5}{7}$

13. $\frac{1}{4}$
     $+ \frac{7}{8}$

14. $\frac{2}{3}$
     $+ \frac{3}{15}$

15. $\frac{2}{6}$
     $+ \frac{1}{3}$

Adding Fractions with Unlike Denominators

Solve each problem. Write each answer in simplest form.

1. $\frac{7}{8}$ $+\frac{1}{4}$ 2. $\frac{1}{3}$ $+\frac{5}{6}$ 3. $\frac{1}{12}$ $+\frac{1}{10}$ 4. $\frac{2}{7}$ $+\frac{1}{5}$ 5. $\frac{3}{10}$ $+\frac{4}{5}$

6. $\frac{1}{12}$ $+\frac{3}{4}$ 7. $\frac{2}{5}$ $+\frac{5}{10}$ 8. $\frac{4}{5}$ $+\frac{3}{6}$ 9. $\frac{1}{4}$ $+\frac{1}{2}$ 10. $\frac{2}{3}$ $+\frac{4}{9}$

11. $\frac{1}{8}$ $+\frac{5}{9}$ 12. $\frac{2}{7}$ $+\frac{1}{3}$ 13. $\frac{1}{10}$ $+\frac{4}{8}$ 14. $\frac{5}{8}$ $+\frac{1}{2}$ 15. $\frac{2}{3}$ $+\frac{1}{6}$

16. $\frac{4}{8}$ $+\frac{3}{7}$ 17. $\frac{2}{3}$ $+\frac{5}{6}$ 18. $\frac{5}{12}$ $+\frac{1}{4}$ 19. $\frac{6}{12}$ $+\frac{5}{6}$ 20. $\frac{1}{3}$ $+\frac{3}{4}$

Adding Fractions with Unlike Denominators

Solve each problem. Write each answer in simplest form.

1. $\frac{2}{5}$
    $+\frac{1}{2}$

2. $\frac{2}{3}$
    $+\frac{3}{4}$

3. $\frac{6}{12}$
    $+\frac{3}{10}$

4. $\frac{2}{6}$
    $+\frac{1}{8}$

5. $\frac{3}{10}$
    $+\frac{1}{3}$

6. $\frac{3}{12}$
    $+\frac{17}{36}$

7. $\frac{2}{6}$
    $+\frac{5}{12}$

8. $\frac{4}{8}$
    $+\frac{3}{5}$

9. $\frac{7}{8}$
    $+\frac{1}{3}$

10. $\frac{5}{6}$
    $+\frac{2}{5}$

11. $\frac{1}{7}$
    $+\frac{5}{8}$

12. $\frac{2}{9}$
    $+\frac{2}{3}$

13. $\frac{2}{10}$
    $+\frac{3}{4}$

14. $\frac{5}{6}$
    $+\frac{1}{4}$

15. $\frac{2}{12}$
    $+\frac{2}{5}$

16. $\frac{2}{4}$
    $+\frac{3}{7}$

17. $\frac{12}{14}$
    $+\frac{4}{5}$

18. $\frac{3}{12}$
    $+\frac{2}{14}$

19. $\frac{5}{13}$
    $+\frac{2}{4}$

20. $\frac{1}{13}$
    $+\frac{3}{6}$

Adding Mixed Numbers with Unlike Denominators

1. Find the least common denominator and equivalent fractions.

$3\frac{2}{3}$ $\frac{2 \times 3}{3 \times 3} = \frac{6}{9}$

$+2\frac{7}{9}$ $\frac{7 \times 1}{9 \times 1} = \frac{7}{9}$

2. Add.

$3\frac{6}{9}$

$+2\frac{7}{9}$

$5\frac{13}{9}$

3. Reduce and regroup if necessary.

$3\frac{6}{9}$

$+2\frac{7}{9}$

$5\frac{13}{9} = 6\frac{4}{9}$

Solve each problem. Write each answer in simplest form.

1. $4\frac{5}{8}$
$+3\frac{1}{6}$

2. $2\frac{5}{6}$
$+6\frac{3}{4}$

3. $4\frac{5}{8}$
$+5\frac{4}{12}$

4. $10\frac{3}{8}$
$+3\frac{1}{2}$

5. $3\frac{2}{5}$
$+2\frac{1}{2}$

6. $8\frac{5}{7}$
$+9\frac{2}{3}$

7. $8\frac{2}{3}$
$+1\frac{5}{9}$

8. $2\frac{3}{4}$
$+7\frac{1}{2}$

9. $1\frac{7}{9}$
$+4\frac{1}{5}$

10. $6\frac{5}{6}$
$+2\frac{2}{3}$

11. $4\frac{2}{14}$
$+6\frac{3}{7}$

12. $1\frac{1}{4}$
$+5\frac{10}{12}$

Name _____

Adding Mixed Numbers with Unlike Denominators

Solve each problem. Write each answer in simplest form.

1. $4\frac{1}{8} + 5\frac{3}{4} =$

2. $4\frac{7}{8} + 6\frac{1}{4} =$

3. $4\frac{3}{4} + 1\frac{2}{3} =$

4. $4\frac{1}{8} + 5\frac{1}{5} =$

5. $8\frac{3}{4} + 7\frac{3}{16} =$

6. $6\frac{1}{2} + 6\frac{2}{5} =$

7. $8\frac{1}{3} + 2\frac{3}{7} =$

8. $5\frac{1}{8} + 6\frac{2}{5} =$

9. $1\frac{9}{10} + 3\frac{1}{4} =$

10. $2\frac{3}{4} + 3\frac{5}{6} =$

11. $3\frac{1}{9} + 2\frac{1}{3} =$

12. $5\frac{2}{3} + 7\frac{3}{7} =$

13. $2\frac{5}{6} + 3\frac{1}{3} =$

14. $5\frac{1}{2} + 6\frac{2}{7} =$

15. $5\frac{5}{15} + 2\frac{3}{5} =$

© Carson-Dellosa • CD-104630

Name _____

Adding Mixed Numbers with Unlike Denominators

Solve each problem. Write each answer in simplest form.

1. $1\frac{1}{4}$ $+ 2\frac{5}{6}$

2. $6\frac{7}{12}$ $+ 6\frac{7}{13}$

3. $5\frac{10}{11}$ $+ 6\frac{3}{22}$

4. $10\frac{4}{5}$ $+ 7\frac{1}{8}$

5. $3\frac{3}{4}$ $+ 6\frac{1}{3}$

6. $9\frac{3}{5}$ $+ 6\frac{2}{3}$

7. $5\frac{1}{2}$ $+ 6\frac{1}{5}$

8. $3\frac{2}{3}$ $+ 4\frac{4}{5}$

9. $8\frac{11}{17}$ $+ 6\frac{2}{3}$

10. $2\frac{1}{15}$ $+ 6\frac{1}{14}$

11. $8\frac{5}{6}$ $+ 3\frac{4}{7}$

12. $5\frac{1}{3}$ $+ 2\frac{3}{4}$

13. $4\frac{1}{8}$ $+ 3\frac{1}{12}$

14. $8\frac{9}{10}$ $+ 4\frac{1}{4}$

15. $8\frac{1}{16}$ $+ 3\frac{3}{4}$

Subtracting Fractions with Unlike Denominators

1. Find the least common denominator (LCD).

$$\frac{3}{4} - \frac{2}{5}$$

LCD = 20

2. Find the equivalent fractions.

$$\frac{3}{4} \times \frac{5}{5} = \frac{15}{20}$$

$$\frac{2}{5} \times \frac{4}{4} = \frac{8}{20}$$

3. Subtract. Reduce to simplest form if necessary.

$$\begin{array}{r} \frac{15}{20} \\ -\frac{8}{20} \\ \hline \frac{7}{20} \end{array}$$

Solve each problem. Write each answer in simplest form.

1. $\frac{1}{3}$
    $-\frac{1}{4}$

2. $\frac{3}{4}$
    $-\frac{1}{5}$

3. $\frac{9}{10}$
    $-\frac{5}{7}$

4. $\frac{5}{7}$
    $-\frac{2}{9}$

5. $\frac{3}{5}$
    $-\frac{1}{3}$

6. $\frac{3}{8}$
    $-\frac{2}{6}$

7. $\frac{2}{4}$
    $-\frac{1}{3}$

8. $\frac{1}{5}$
    $-\frac{1}{8}$

9. $\frac{7}{12}$
    $-\frac{1}{4}$

10. $\frac{3}{9}$
    $-\frac{1}{4}$

11. $\frac{7}{8}$
    $-\frac{1}{9}$

12. $\frac{8}{8}$
    $-\frac{4}{6}$

13. $\frac{2}{3}$
    $-\frac{1}{2}$

14. $\frac{2}{3}$
    $-\frac{4}{9}$

15. $\frac{1}{3}$
    $-\frac{1}{6}$

Subtracting Fractions with Unlike Denominators

Solve each problem. Write each answer in simplest form.

1.
$$\begin{array}{r} \frac{3}{9} \\ -\frac{1}{4} \\ \hline \end{array}$$

2.
$$\begin{array}{r} \frac{3}{10} \\ -\frac{1}{5} \\ \hline \end{array}$$

3.
$$\begin{array}{r} \frac{4}{5} \\ -\frac{5}{10} \\ \hline \end{array}$$

4.
$$\begin{array}{r} \frac{15}{27} \\ -\frac{4}{9} \\ \hline \end{array}$$

5.
$$\begin{array}{r} \frac{2}{3} \\ -\frac{4}{9} \\ \hline \end{array}$$

6.
$$\begin{array}{r} \frac{7}{8} \\ -\frac{2}{16} \\ \hline \end{array}$$

7.
$$\begin{array}{r} \frac{2}{4} \\ -\frac{1}{12} \\ \hline \end{array}$$

8.
$$\begin{array}{r} \frac{10}{15} \\ -\frac{1}{8} \\ \hline \end{array}$$

9.
$$\begin{array}{r} \frac{7}{10} \\ -\frac{2}{4} \\ \hline \end{array}$$

10.
$$\begin{array}{r} \frac{3}{9} \\ -\frac{1}{3} \\ \hline \end{array}$$

11.
$$\begin{array}{r} \frac{7}{8} \\ -\frac{1}{2} \\ \hline \end{array}$$

12.
$$\begin{array}{r} \frac{8}{18} \\ -\frac{4}{9} \\ \hline \end{array}$$

13.
$$\begin{array}{r} \frac{12}{20} \\ -\frac{1}{2} \\ \hline \end{array}$$

14.
$$\begin{array}{r} \frac{1}{2} \\ -\frac{1}{4} \\ \hline \end{array}$$

15.
$$\begin{array}{r} \frac{8}{10} \\ -\frac{1}{6} \\ \hline \end{array}$$

16.
$$\begin{array}{r} \frac{8}{9} \\ -\frac{3}{6} \\ \hline \end{array}$$

17.
$$\begin{array}{r} \frac{5}{6} \\ -\frac{1}{5} \\ \hline \end{array}$$

18.
$$\begin{array}{r} \frac{7}{8} \\ -\frac{3}{10} \\ \hline \end{array}$$

19.
$$\begin{array}{r} \frac{9}{12} \\ -\frac{2}{11} \\ \hline \end{array}$$

20.
$$\begin{array}{r} \frac{6}{6} \\ -\frac{3}{12} \\ \hline \end{array}$$

Subtracting Fractions with Unlike Denominators

Solve each problem. Write each answer in simplest form.

1. $\dfrac{3}{4}$
    $-\dfrac{1}{6}$

2. $\dfrac{13}{15}$
    $-\dfrac{2}{3}$

3. $\dfrac{2}{3}$
    $-\dfrac{7}{12}$

4. $\dfrac{5}{6}$
    $-\dfrac{1}{3}$

5. $\dfrac{5}{6}$
    $-\dfrac{2}{5}$

6. $\dfrac{2}{3}$
    $-\dfrac{1}{6}$

7. $\dfrac{11}{14}$
    $-\dfrac{1}{2}$

8. $\dfrac{7}{12}$
    $-\dfrac{1}{4}$

9. $\dfrac{11}{12}$
    $-\dfrac{1}{16}$

10. $\dfrac{5}{6}$
    $-\dfrac{3}{7}$

11. $\dfrac{7}{8}$
    $-\dfrac{1}{9}$

12. $\dfrac{7}{8}$
    $-\dfrac{1}{12}$

13. $\dfrac{5}{12}$
    $-\dfrac{1}{13}$

14. $\dfrac{7}{8}$
    $-\dfrac{1}{6}$

15. $\dfrac{1}{3}$
    $-\dfrac{1}{6}$

16. $\dfrac{2}{3}$
    $-\dfrac{4}{9}$

17. $\dfrac{3}{5}$
    $-\dfrac{1}{13}$

18. $\dfrac{8}{9}$
    $-\dfrac{5}{6}$

19. $\dfrac{9}{12}$
    $-\dfrac{2}{11}$

20. $\dfrac{5}{6}$
    $-\dfrac{1}{8}$

Subtracting Mixed Numbers with Unlike Denominators

1. Find the least common denominator and equivalent fractions.

$5\frac{1}{8}$ $\frac{1 \times 3}{8 \times 3} = \frac{3}{24}$

$-2\frac{1}{3}$ $\frac{1 \times 8}{3 \times 8} = \frac{8}{24}$

2. Borrow and regroup. Subtract the fractions.

$5\frac{3}{24}$

$-2\frac{8}{24}$

$4\frac{27}{24}$ (from $\cancel{5}\frac{\cancel{3}}{24}$)

$-2\frac{8}{24}$

$\frac{19}{24}$

3. Subtract the whole numbers.

$4\cancel{5}\frac{\cancel{3}}{24}$ ($\frac{27}{24}$)

$-2\frac{8}{24}$

$2\frac{19}{24}$

Reduce to lowest terms if necessary.

Solve each problem. Write each answer in simplest form.

1. $5\frac{1}{6}$
    $-2\frac{3}{4}$

2. $4\frac{7}{10}$
    $-1\frac{4}{5}$

3. $5\frac{7}{8}$
    $-1\frac{1}{16}$

4. $3\frac{1}{3}$
    $-5\frac{5}{6}$

5. $4\frac{1}{3}$
    $-1\frac{1}{4}$

6. $3\frac{7}{12}$
    $-1\frac{9}{10}$

7. $5\frac{4}{5}$
    $-1\frac{9}{10}$

8. $4\frac{3}{4}$
    $-1\frac{5}{6}$

9. $6\frac{1}{2}$
    $-\frac{1}{3}$

10. $7\frac{1}{4}$
     $-3\frac{2}{3}$

11. $10\frac{4}{5}$
     $-6\frac{5}{6}$

12. $12\frac{2}{3}$
     $-9\frac{6}{7}$

Subtracting Mixed Numbers with Unlike Denominators

Solve each problem. Write each answer in simplest form.

1. $12\frac{7}{8}$
$-5\frac{5}{16}$

2. $3\frac{1}{4}$
$-2\frac{5}{12}$

3. $10\frac{2}{3}$
$-9\frac{2}{9}$

4. $3\frac{1}{8}$
$-1\frac{7}{9}$

5. $10\frac{2}{5}$
$-7\frac{2}{3}$

6. $8\frac{7}{10}$
$-7\frac{9}{11}$

7. $8\frac{5}{10}$
$-7\frac{5}{12}$

8. $5\frac{12}{16}$
$-5\frac{11}{20}$

9. $6\frac{1}{6}$
$-5\frac{5}{12}$

10. $4\frac{5}{6}$
$-2\frac{1}{24}$

11. $8\frac{3}{16}$
$-7\frac{5}{32}$

12. $6\frac{1}{9}$
$-2\frac{1}{3}$

13. $2\frac{1}{3}$
$-1\frac{1}{5}$

14. $9\frac{3}{5}$
$-4\frac{19}{20}$

15. $4\frac{11}{18}$
$-1\frac{13}{16}$

16. $8\frac{7}{10}$
$-6\frac{3}{40}$

Subtracting Mixed Numbers with Unlike Denominators

Solve each problem. Write each answer in simplest form.

1. $5\frac{4}{9}$
$-\ 2\frac{1}{3}$

2. $3\frac{1}{2}$
$-\ 2\frac{3}{4}$

3. $9\frac{1}{3}$
$-\ 1\frac{2}{5}$

4. $5\frac{5}{12}$
$-\ 3\frac{7}{10}$

5. $3\frac{5}{6}$
$-\ 1\frac{5}{9}$

6. $7\frac{3}{5}$
$-\ 4\frac{7}{10}$

7. $6\frac{1}{4}$
$-\ 4\frac{1}{2}$

8. $4\frac{7}{8}$
$-\ 2\frac{1}{4}$

9. $4\frac{2}{5}$
$-\ 2\frac{3}{10}$

10. $6\frac{4}{5}$
$-\ 5\frac{3}{7}$

11. 7
$-\ \frac{5}{6}$

12. 2
$-\ \frac{4}{5}$

13. 2
$-\ \frac{6}{11}$

14. 1
$-\ \frac{7}{8}$

15. 5
$-\ \frac{1}{4}$

16. 6
$-\ \frac{6}{9}$

17. 5
$-\ \frac{3}{5}$

18. 10
$-\ \frac{1}{3}$

19. 8
$-\ \frac{3}{4}$

20. 7
$-\ \frac{3}{7}$

Understanding Fractions as Division

Each fraction is written as a division problem. Solve.

1. $\dfrac{15}{8} = 8\overline{)15}$

2. $\dfrac{25}{13} = 13\overline{)25}$

3. $\dfrac{54}{7} = 7\overline{)54}$

4. $\dfrac{29}{5} = 5\overline{)29}$

5. $\dfrac{85}{22} = 22\overline{)85}$

6. $\dfrac{10}{4} = 4\overline{)10}$

7. $\dfrac{62}{8} = 8\overline{)62}$

8. $\dfrac{34}{10} = 10\overline{)34}$

9. $\dfrac{43}{16} = 16\overline{)43}$

10. $\dfrac{33}{7} = 7\overline{)33}$

Understanding Fractions as Division

Write each fraction as a division problem. Solve.

1. $\frac{16}{7} =$

2. $\frac{32}{9} =$

3. $\frac{4}{3} =$

4. $\frac{37}{12} =$

5. $\frac{12}{5} =$

6. $\frac{20}{15} =$

7. $\frac{50}{27} =$

8. $\frac{10}{4} =$

9. $\frac{7}{4} =$

10. $\frac{53}{16} =$

11. $\frac{86}{19} =$

12. $\frac{55}{12} =$

13. $\frac{14}{13} =$

14. $\frac{43}{20} =$

15. $\frac{18}{5} =$

Understanding Fractions as Division

Write each fraction as a division problem. Solve.

1. $\dfrac{6}{4}=$

2. $\dfrac{21}{12}=$

3. $\dfrac{9}{4}=$

4. $\dfrac{5}{11}=$

5. $\dfrac{19}{5}=$

6. $\dfrac{3}{2}=$

7. $\dfrac{7}{4}=$

8. $\dfrac{13}{3}=$

9. $\dfrac{14}{6}=$

10. $\dfrac{16}{5}=$

11. $\dfrac{3}{5}=$

12. $\dfrac{14}{8}=$

13. $\dfrac{11}{2}=$

14. $\dfrac{17}{4}=$

15. $\dfrac{19}{2}=$

16. $\dfrac{2}{3}=$

17. $\dfrac{8}{3}=$

18. $\dfrac{1}{6}=$

19. $\dfrac{1}{3}=$

20. $\dfrac{3}{6}=$

21. $\dfrac{14}{9}=$

22. $\dfrac{21}{8}=$

23. $\dfrac{4}{7}=$

24. $\dfrac{13}{3}=$

25. $\dfrac{12}{5}=$

26. $\dfrac{8}{11}=$

27. $\dfrac{9}{2}=$

28. $\dfrac{15}{4}=$

29. $\dfrac{10}{6}=$

30. $\dfrac{1}{4}=$

Multiplying Fractions

1. Multiply the numerators.	2. Multiply the denominators.	3. Simplify when necessary.
$\dfrac{2}{3} \times \dfrac{5}{6} = \dfrac{10}{}$	$\dfrac{2}{3} \times \dfrac{5}{6} = \dfrac{10}{18}$	$\dfrac{10}{18} \div \dfrac{2}{2} = \dfrac{5}{9}$

Solve each problem. Write the answer in simplest form.

1. $\dfrac{3}{4} \times \dfrac{2}{5} =$

2. $\dfrac{7}{8} \times \dfrac{1}{6} =$

3. $\dfrac{4}{5} \times \dfrac{2}{3} =$

4. $\dfrac{1}{3} \times \dfrac{1}{5} =$

5. $\dfrac{2}{7} \times \dfrac{2}{9} =$

6. $\dfrac{1}{4} \times \dfrac{3}{5} =$

7. $\dfrac{4}{7} \times \dfrac{3}{8} =$

8. $\dfrac{2}{3} \times \dfrac{2}{5} =$

9. $\dfrac{1}{3} \times \dfrac{3}{5} =$

10. $\dfrac{3}{5} \times \dfrac{1}{3} =$

11. $\dfrac{1}{8} \times \dfrac{2}{5} =$

12. $\dfrac{1}{6} \times \dfrac{2}{3} =$

Multiplying Fractions

Solve each problem. Write each answer in simplest form.

1. $\frac{1}{3} \times \frac{1}{7} =$

2. $\frac{3}{5} \times \frac{2}{9} =$

3. $\frac{1}{6} \times \frac{4}{5} =$

4. $\frac{2}{7} \times \frac{5}{8} =$

5. $\frac{2}{5} \times \frac{4}{9} =$

6. $\frac{1}{4} \times \frac{1}{6} =$

7. $\frac{2}{3} \times \frac{3}{8} =$

8. $\frac{3}{4} \times \frac{4}{7} =$

9. $\frac{2}{5} \times \frac{5}{6} =$

10. $\frac{4}{5} \times \frac{2}{3} =$

11. $\frac{1}{5} \times \frac{5}{6} =$

12. $\frac{1}{2} \times \frac{3}{7} =$

13. $\frac{2}{5} \times \frac{4}{9} =$

14. $\frac{2}{8} \times \frac{3}{3} =$

15. $\frac{1}{7} \times \frac{6}{8} =$

Multiplying Fractions

Solve each problem. Write each answer in simplest form.

1. $\dfrac{1}{4} \times \dfrac{2}{5} =$

2. $\dfrac{2}{8} \times \dfrac{3}{6} =$

3. $\dfrac{1}{6} \times \dfrac{4}{5} =$

4. $\dfrac{1}{3} \times \dfrac{5}{6} =$

5. $\dfrac{4}{6} \times \dfrac{5}{7} =$

6. $\dfrac{3}{5} \times \dfrac{1}{8} =$

7. $\dfrac{5}{7} \times \dfrac{2}{4} =$

8. $\dfrac{3}{4} \times \dfrac{4}{7} =$

9. $\dfrac{3}{4} \times \dfrac{5}{15} =$

10. $\dfrac{12}{16} \times \dfrac{4}{5} =$

11. $\dfrac{5}{6} \times \dfrac{3}{4} =$

12. $\dfrac{3}{5} \times \dfrac{3}{7} =$

13. $\dfrac{2}{5} \times \dfrac{4}{9} =$

14. $\dfrac{12}{18} \times \dfrac{3}{13} =$

15. $\dfrac{1}{7} \times \dfrac{3}{4} =$

Multiplying Whole Numbers and Fractions

1. Convert the whole number to a fraction.

$$7 \times \frac{2}{3} = \frac{7}{1} \times \frac{2}{3}$$

2. Multiply straight across.

$$\frac{7}{1} \times \frac{2}{3} = \frac{14}{3}$$

3. If the product is an improper fraction, convert to a mixed number in simplest form.

$$\frac{14}{3} = 4\frac{2}{3}$$

Solve each problem. Write each answer in simplest form.

1. $5 \times \frac{2}{5} =$

2. $8 \times \frac{1}{7} =$

3. $6 \times \frac{3}{8} =$

4. $4 \times \frac{8}{9} =$

5. $2 \times \frac{3}{7} =$

6. $\frac{2}{3} \times 4 =$

7. $\frac{1}{9} \times 6 =$

8. $\frac{5}{6} \times 4 =$

9. $\frac{4}{6} \times 3 =$

10. $\frac{4}{5} \times 6 =$

11. $\frac{3}{4} \times 5 =$

12. $2 \times \frac{4}{5} =$

Multiplying Whole Numbers and Fractions

Solve **each** problem. Write each answer in simplest form.

1. $10 \times \frac{2}{3} =$

2. $4 \times \frac{4}{7} =$

3. $7 \times \frac{10}{11} =$

4. $36 \times \frac{2}{288} =$

5. $6 \times \frac{4}{8} =$

6. $9 \times \frac{5}{6} =$

7. $3 \times \frac{1}{3} =$

8. $30 \times \frac{3}{90} =$

9. $12 \times \frac{1}{36} =$

10. $5 \times \frac{2}{5} =$

11. $12 \times \frac{7}{8} =$

12. $5 \times \frac{3}{4} =$

13. $22 \times \frac{1}{44} =$

14. $4 \times \frac{1}{8} =$

15. $8 \times \frac{2}{3} =$

Name _____

Multiplying Whole Numbers and Fractions

Solve each problem. Write each answer in simplest form.

1. $2 \times 2\frac{1}{3} =$

2. $3 \times 5\frac{1}{5} =$

3. $9 \times 3\frac{2}{3} =$

4. $8 \times 9\frac{1}{10} =$

5. $4 \times 5\frac{1}{8} =$

6. $6 \times 3\frac{1}{6} =$

7. $5 \times 6\frac{5}{8} =$

8. $3 \times 9\frac{1}{3} =$

9. $7 \times 1\frac{3}{4} =$

10. $7 \times 2\frac{3}{5} =$

11. $4 \times 2\frac{1}{2} =$

12. $7 \times 2\frac{1}{7} =$

13. $3 \times 1\frac{15}{16} =$

14. $4 \times 8\frac{6}{7} =$

15. $2 \times 2\frac{1}{4} =$

Multiplying Mixed Numbers

1. Convert the mixed numbers to fractions.

$$1\frac{1}{3} \times 2\frac{1}{2} = \frac{4}{3} \times \frac{2}{2}$$

2. Multiply.

$$\frac{4}{3} \times \frac{2}{2} = \frac{20}{6}$$

3. Convert the product back to a mixed number in simplest form.

$$\frac{20}{6} = 3\frac{2}{6} = 3\frac{1}{3}$$

Solve each problem. Write each answer in simplest form.

1. $8\frac{1}{4} \times 6\frac{2}{3} =$

2. $7\frac{2}{5} \times 6\frac{2}{3} =$

3. $2\frac{5}{6} \times 12\frac{4}{5} =$

4. $4\frac{2}{7} \times 6\frac{1}{10} =$

5. $5\frac{1}{5} \times 4\frac{1}{3} =$

6. $9\frac{9}{10} \times 4\frac{7}{8} =$

7. $1\frac{10}{13} \times 2\frac{9}{13} =$

8. $8\frac{3}{5} \times 4\frac{5}{6} =$

Multiplying Mixed Numbers

Solve each problem. Write each answer in simplest form.

1. $3\frac{1}{2} \times 2\frac{1}{2} =$

2. $8\frac{5}{6} \times 3\frac{6}{7} =$

3. $4\frac{2}{5} \times 6\frac{2}{3} =$

4. $4\frac{2}{9} \times 5\frac{10}{11} =$

5. $2\frac{2}{3} \times 4\frac{2}{5} =$

6. $5\frac{3}{4} \times 6\frac{1}{4} =$

7. $2\frac{8}{9} \times 7\frac{7}{8} =$

8. $7\frac{1}{4} \times 3\frac{3}{7} =$

9. $6\frac{7}{8} \times 3\frac{1}{3} =$

10. $7\frac{9}{10} \times 8\frac{7}{8} =$

11. $4\frac{1}{4} \times 3\frac{5}{6} =$

12. $8\frac{3}{5} \times 1\frac{1}{2} =$

Multiplying Mixed Numbers

Solve each problem. Write each answer in simplest form.

1. $8\frac{1}{4} \times 6\frac{2}{3} =$

2. $7\frac{2}{5} \times 9\frac{1}{8} =$

3. $12\frac{4}{6} \times 2\frac{4}{15} =$

4. $4\frac{2}{17} \times 6\frac{9}{10} =$

5. $2\frac{9}{10} \times 5\frac{7}{8} =$

6. $5\frac{1}{3} \times 4\frac{1}{2} =$

7. $3\frac{1}{3} \times 3\frac{1}{3} =$

8. $3\frac{3}{4} \times 2\frac{1}{3} =$

9. $5\frac{10}{15} \times 3\frac{2}{3} =$

10. $9\frac{9}{10} \times 4\frac{7}{8} =$

11. $11\frac{1}{3} \times 4\frac{7}{13} =$

12. $10\frac{8}{15} \times 4\frac{2}{6} =$

Dividing Whole Numbers by Unit Fractions

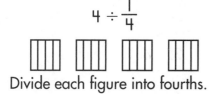

$4 \div \frac{1}{4}$

Divide each figure into fourths.

$4 \div \frac{1}{4} = 16$ There are now 16 parts.

1. To solve, turn the division problem into a multiplication problem by flipping the digits in the fraction.

$4 \times \frac{4}{1} = 16$

2. Check the quotient by multiplying it by the divisor.

$16 \times \frac{1}{4} = 4$

$\frac{16}{4} = 4$

Solve each problem. Write each answer in simplest form.

1. $1 \div \frac{1}{4} =$

2. $3 \div \frac{1}{8} =$

3. $5 \div \frac{1}{10} =$

4. $1 \div \frac{1}{7} =$

5. $2 \div \frac{1}{8} =$

6. $2 \div \frac{1}{2} =$

7. $2 \div \frac{1}{8} =$

8. $4 \div \frac{1}{2} =$

9. $6 \div \frac{1}{7} =$

Dividing Whole Numbers by Unit Fractions

Solve each problem. Write each answer in simplest form.

1. $1 \div \frac{1}{4} =$

2. $2 \div \frac{1}{3} =$

3. $6 \div \frac{1}{10} =$

4. $1 \div \frac{1}{8} =$

5. $3 \div \frac{1}{4} =$

6. $5 \div \frac{1}{8} =$

7. $2 \div \frac{1}{6} =$

8. $2 \div \frac{1}{5} =$

9. $1 \div \frac{1}{3} =$

10. $3 \div \frac{1}{10} =$

11. $1 \div \frac{1}{5} =$

12. $5 \div \frac{1}{4} =$

Dividing Whole Numbers by Unit Fractions

Solve each problem. Write each answer in simplest form.

1. $1 \div \frac{1}{2} =$

2. $6 \div \frac{1}{3} =$

3. $3 \div \frac{1}{10} =$

4. $1 \div \frac{1}{5} =$

5. $11 \div \frac{1}{5} =$

6. $2 \div \frac{1}{2} =$

7. $2 \div \frac{1}{3} =$

8. $4 \div \frac{1}{18} =$

9. $12 \div \frac{1}{2} =$

10. $4 \div \frac{1}{5} =$

11. $3 \div \frac{1}{6} =$

12. $3 \div \frac{1}{8} =$

13. $8 \div \frac{1}{12} =$

14. $5 \div \frac{1}{9} =$

15. $9 \div \frac{1}{15} =$

Dividing Unit Fractions by Whole Numbers

1. Change the whole number to a fraction.

$$\frac{1}{3} \div 6 = \frac{1}{3} \div \frac{6}{1}$$

2. Find the reciprocal of the second fraction by flipping it. Change the division sign to a multiplication sign.

$$\frac{6}{1} = \frac{1}{6}$$

$$\frac{1}{3} \div 6 = \frac{1}{3} \times \frac{1}{6}$$

3. Multiply. Simplify when necessary.

$$\frac{1}{3} \times \frac{1}{6} = \frac{1}{18}$$

Solve each problem. Write each answer in simplest form.

1. $\frac{1}{3} \div 4 =$

2. $\frac{1}{2} \div 1 =$

3. $\frac{1}{3} \div 1 =$

4. $\frac{1}{3} \div 2 =$

5. $\frac{1}{4} \div 2 =$

6. $\frac{1}{8} \div 1 =$

7. $\frac{1}{3} \div 8 =$

8. $\frac{1}{5} \div 3 =$

9. $\frac{1}{3} \div 3 =$

Dividing Unit Fractions by Whole Numbers

Solve each problem. Write each answer in simplest form.

1. $\frac{1}{4} \div 5 =$

2. $\frac{1}{5} \div 8 =$

3. $\frac{1}{3} \div 5 =$

4. $\frac{1}{7} \div 2 =$

5. $\frac{1}{5} \div 2 =$

6. $\frac{1}{5} \div 1 =$

7. $\frac{1}{2} \div 3 =$

8. $\frac{1}{3} \div 15 =$

9. $\frac{1}{5} \div 3 =$

10. $\frac{1}{5} \div 6 =$

11. $\frac{1}{4} \div 2 =$

12. $\frac{1}{8} \div 3 =$

Dividing Unit Fractions by Whole Numbers

Solve each problem. Write each answer in simplest form.

1. $\frac{1}{3} \div 1 =$

2. $\frac{1}{2} \div 1 =$

3. $\frac{1}{6} \div 1 =$

4. $\frac{1}{6} \div 4 =$

5. $\frac{1}{5} \div 2 =$

6. $\frac{1}{4} \div 1 =$

7. $\frac{1}{4} \div 11 =$

8. $\frac{1}{8} \div 4 =$

9. $\frac{1}{2} \div 3 =$

10. $\frac{1}{3} \div 2 =$

11. $\frac{1}{5} \div 11 =$

12. $\frac{1}{5} \div 3 =$

13. $\frac{1}{9} \div 8 =$

14. $\frac{1}{7} \div 4 =$

15. $\frac{1}{12} \div 7 =$

Fraction Word Problems

Create a story problem that makes sense with each problem. Then, solve.

1. $\frac{4}{6} + \frac{7}{8}$

_____

_____

_____

_____

2. $5\frac{6}{9} - \frac{1}{5}$

_____

_____

_____

_____

3. $4\frac{3}{4} + 1\frac{1}{3}$

_____

_____

_____

_____

4. $6 - 1\frac{2}{5}$

_____

_____

_____

_____

Fraction Word Problems

Create a story problem that makes sense with each problem. Then, solve.

1. $\frac{1}{4} \times \frac{3}{5}$

_____

_____

_____

_____

2. $\frac{5}{11} \times \frac{2}{3}$

_____

_____

_____

_____

3. $1\frac{4}{5} \times \frac{4}{5}$

_____

_____

_____

_____

4. $8 \times \frac{5}{8}$

_____

_____

_____

_____

Fraction Word Problems

Create a story problem that makes sense with each problem. Then, solve.

1. $10 \div \frac{1}{4}$

_____

_____

_____

_____

2. $7 \div \frac{1}{3}$

_____

_____

_____

_____

3. $\frac{1}{6} \div 9$

_____

_____

_____

_____

4. $\frac{1}{5} \div 4$

_____

_____

_____

_____

Converting Measurements

Standard units of length

12 inches (in.) = 1 foot (ft.)

3 feet (ft.) = 1 yard (yd.)

5,280 feet (ft.) = 1 mile (mi.)

1,760 yards (yd.) = 1 mile (mi.)

Metric units of length

10 millimeters (mm) = 1 centimeter (cm)

100 centimeters (cm) = 1 meter (m)

1,000 meters (m) = 1 kilometer (km)

US standard units of capacity and weight

2 cups (c.) = 1 pint (pt.)

2 pints = 1 quart (qt.)

4 quarts = 1 gallon (gal.)

16 ounces (oz.) = 1 pound (lb.)

2,000 pounds = 1 ton (t.)

Metric units of capacity and weight

1,000 milliliters (mL) = 1 liter (L)

1,000 liters (L) = 1 kiloliter (kL)

1,000 milligrams (mg) = 1 gram (g)

1,000 grams (g) = 1 kilogram (kg)

Convert each unit of weight.

1. 32 oz. = _____ lb.

2. 3 lb. = _____ oz.

3. 5 kg = _____ g

4. 8 oz. = _____ lb.

5. 4 g = _____ kg

6. 4 oz. = _____ lb.

7. 6,000 mg = _____ g

8. 1 t. = _____ lb.

9. 4,000 lb. = _____ t.

10. 80 oz. = _____ lb.

11. 800 g = _____ kg

12. 6 lb. = _____ oz.

Convert each unit of capcity.

13. 3,000 mL = _____ L

14. 3 c. = _____ pt.

15. 4 L = _____ mL

16. 3 pt. = _____ c.

17. 4 qt. = _____ gal.

18. 4 pt. = _____ qt.

19. 2 kL = _____ L

20. 1 qt. = _____ gal.

21. 1 pt.. = _____ qt.

Convert each unit of length.

22. 20 mm = _____ cm

23. 5 yd. = _____ in.

24. 3 mi. = _____ ft.

25. 50 m = _____ cm

26. 2 m = _____ cm

27. 1 mi. = _____ in.

28. 90 cm = _____ mm

29. 10,000 km = _____ m

30. 8 ft. = _____ in.

Converting Measurements

Convert.

1. 1,000 mL = _____ kL

2. 40 g = _____ mg

3. 14 qt. = _____ gal.

4. 12 gal. = _____ c.

5. 9 L = _____ mL

6. 8 lb. = _____ oz.

7. 8 kL = _____ mL

8. 600 mg = _____ g

9. 15 tons = _____ lb.

10. 11,000 L = _____ kL

11. 7 g = _____ mg

12. 14 qt. = _____ pt.

13. 200 mm = _____ m

14. 3 mi. = _____ yd.

15. 12 yd. = _____ in.

16. 30 km = _____ cm

17. 3 m = _____ mm

18. 8,000 cm = _____ km

19. 15 yd. 4 ft. = _____ in.

20. 7,000 m = _____ km

21. 48 in. = _____ ft.

22. 430 yd. = _____ ft.

23. 670 mm = _____ cm

24. 1.5 mi. = _____ ft.

Converting Measurements

Convert.

Length	Capacity	Weight
1. 3 ft. = _____ in.	12. 4 c. = _____ pt.	23. 24 oz. = _____ lb.
2. 48 in. = _____ ft.	13. 4,600 mL = _____ L	24. 4,000 lb. = _____ t.
3. 1,500 cm = _____ m	14. 8 qt. = _____ gal.	25. 17,000 g = _____ kg
4. 4 yd. = _____ ft.	15. 1.5 L = _____ mL	26. 3 kg 300 g = _____ g
5. 15 ft. = _____ yd.	16. 7 pt. = _____ qt.	27. 2 lb. = _____ oz.
6. 300 mm = _____ m	17. 3 gal. = _____ qt.	28. 36,000 g = _____ kg
7. 4.5 km = _____ m	18. 15 kL 200 L = _____ L	29. 3.5 t. = _____ lb.
8. 2.5 mi. = _____ yd.	19. 2 qt. = _____ c.	30. 5 lb. = _____ oz.
9. 6.5 yd. = _____ in.	20. 3 pt. = _____ c.	31. 8.5 kg = _____ g
10. 60 in. = _____ ft.	21. 2,000,000 mL = _____ kL	32. 3 lb. = _____ oz.
11. 1.2 km = _____ mm	22. 4.5 gal. = _____ qt.	33. 10,000 lb. = _____ t.

Use <, >, or = to compare.

34. 8,000 mg ◯ 10 g   35. 7 g ◯ 60 mg   36. 800 mg ◯ 8 g

37. 10 kg ◯ 10,000 g   38. 5 g ◯ 10 mg   39. 18 mg ◯ 20 g

40. 100 mm ◯ 10 cm   41. 10 km ◯ 20 m   42. 3 cm ◯ 35 mm

43. 5 cm ◯ 10 mm   44. 1 km ◯ 300 m   45. 2 km ◯ 2,000 m

Reading Line Plots

Use the line plot to answer the questions.

Length of Sticks in Inches

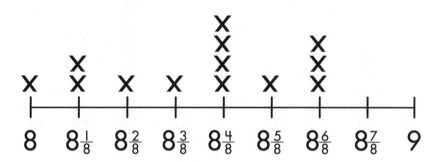

1. How many sticks were part of the data set?

2. How many sticks were shorter than 8.75 inches?

3. Maya needs sticks from 8.25 inches long to 8.75 inches long for a craft project. What fraction of the set can she use?

4. Maya found 8 more sticks with the following measurements:

$$8\frac{1}{8}, 8\frac{1}{8}, 8, 9, 8\frac{5}{8}, 8\frac{3}{8}, 8\frac{5}{8}, 8\frac{2}{8}$$

Add the new data to the line plot. What fraction of the set can Maya use now?

Reading Line Plots

Use the line plot to answer the questions.

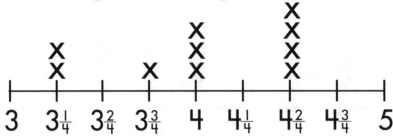

Length of String in Inches

1. How many pieces of string were 4 inches or longer?

2. What fraction of the pieces are not exactly 3, 4, or 5 inches?

3. If the pieces were all cut from the same piece of string, how long was the original piece of string?

4. Five more pieces of string with the following lengths were cut:

$$3\frac{3}{4}, \ 3, \ 4\frac{1}{4}, \ 4\frac{3}{4}, \ 3\frac{3}{4}$$

Add the new data to the line plot. Could all of the lengths on the line plot have been cut from a 5-foot piece of string? Why or why not?

Reading Line Plots

Use the line plot to answer the questions.

Milk in Glasses in Pints

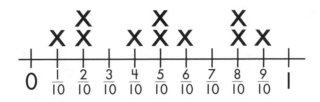

1. How many pints of milk were poured in all?

2. If the milk was poured from a gallon container, how many cups of milk are left in the container?

3. Five more glasses of milk with the following amounts were poured.

$$\frac{3}{10}, \frac{1}{10}, \frac{2}{10}, \frac{3}{10}, \frac{2}{10}$$

Add the new data to the line plot. How much milk has been poured in all now? Did a new gallon of milk need to be opened? Why or why not?

4. Including the new data, how much milk would be in each glass if the total amount in all of the glasses were redistributed equally?

Name _____

Exploring Volume

Volume (V) tells the number of cubic units inside a figure.
Each box represents one cubic unit.

4 cubic units

Write the number of cubic units in each figure.

1.

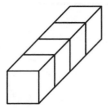

V = _____ cubic units

2.

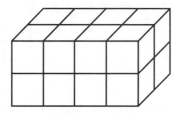

V = _____ cubic units

3.

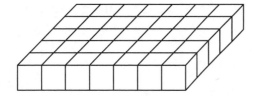

V = _____ cubic units

4.

V = _____ cubic units

5.

V = _____ cubic units

6.

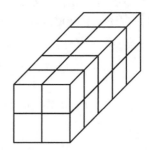

V = _____ cubic units

Exploring Volume

Write the number of cubic units in each figure.

1.

V = _____ cubic units

2.

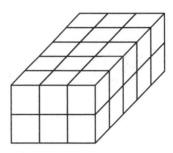

V = _____ cubic units

3.

V = _____ cubic units

4.

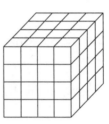

V = _____ cubic units

5.

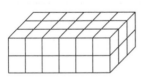

V = _____ cubic units

6.

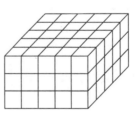

V = _____ cubic units

Exploring Volume

Write the number of cubic units in each figure.

1.

V = _____

2.

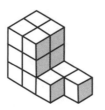

V = _____

3.

V = _____

4.

V = _____

5.

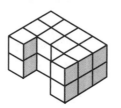

V = _____

6.

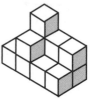

V = _____

Use the figure to the right to answer the questions.

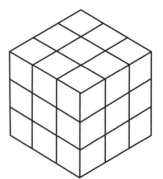

7. What is the volume of this figure?

_____

_____

8. If you painted the front, top, and bottom of the figure, how many cubic units would have one side painted?

_____

_____

9. How many cubic units would have two sides painted?

_____

_____

10. How many cubic units would have no paint on them?

_____

_____

Using a Formula to Find Volume

Find the volume of each figure. Use this formula: Volume = length × width × height.

1. V = _____

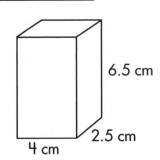

6.5 cm
2.5 cm
4 cm

2. V = _____

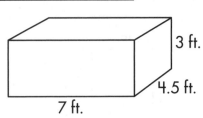

3 ft.
4.5 ft.
7 ft.

3. V = _____

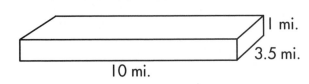

1 mi.
3.5 mi.
10 mi.

4. V = _____

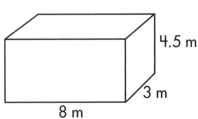

4.7 yd.
2 yd.
5.5 yd.

5. V = _____

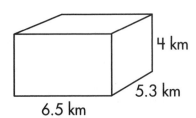

4 km
5.3 km
6.5 km

6. V = _____

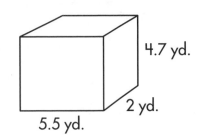

4.5 m
3 m
8 m

Using a Formula to Find Volume

Use the formula ($V = l \times w \times h$) to find the volume of each figure.

1. V = _____

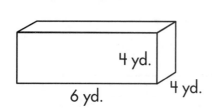

2. V = _____

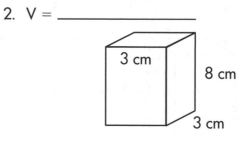

3. V = _____

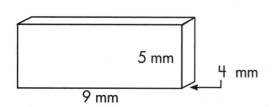

4. V = _____

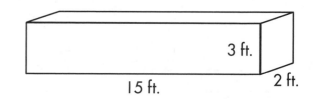

5. V = _____

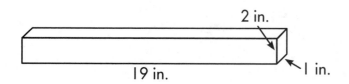

6. V = _____

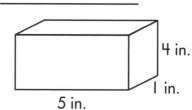

7. V = _____

8. V = _____

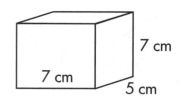

Using a Formula to Find Volume

Use the formula ($V = l \times w \times h$) to find the volume of each rectangular prism.

1. V = _____

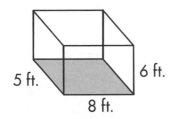

5 ft.   6 ft.
8 ft.

2. V = _____

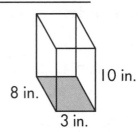

10 in.
8 in.
3 in.

3. V = _____

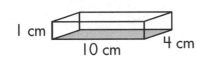

1 cm   4 cm
10 cm

4. V = _____

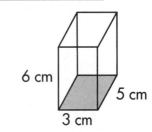

6 cm   5 cm
3 cm

5. V = _____

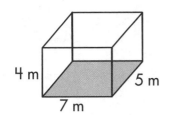

4 m   5 m
7 m

6. V = _____

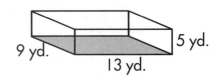

9 yd.   5 yd.
13 yd.

7. V = _____

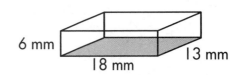

8 m   2 m
16 m

8. V = _____

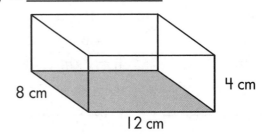

6 mm   13 mm
18 mm

9. V = _____

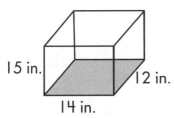

15 in.   12 in.
14 in.

10. V = _____

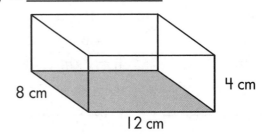

8 cm   4 cm
12 cm

Finding Volume

Use this formula to find volume: $V = l \times w \times h$

Find the volume of each figure.

1. V = _____

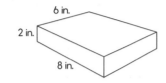

2. V = _____

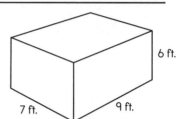

3. V = _____

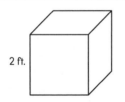

4. V = _____

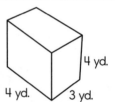

5. V = _____

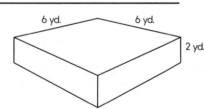

6. V = _____

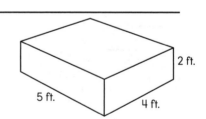

Given the dimensions, find the volume for each rectangular prism.

7. l = 4 cm
    w = 6 cm
    h = 2 cm

8. l = 10 cm
    w = 8 cm
    h = 3 cm

9. l = 9 cm
    w = 4 cm
    h = 12cm

10. l = 5 cm
    w = 5 cm
    h = 7 cm

V = _____ V = _____ V = _____ V = _____

Name _____

Finding Volume

Find the volume of each figure.

1. V = _____ cm³

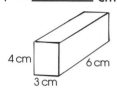

4 cm   6 cm
3 cm

2. V = _____ m³

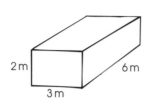

2 m   6 m
3 m

3. V = _____ m³

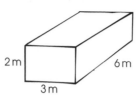

2 m   6 m
3 m

4. V = _____ in.³

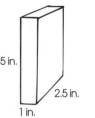

5 in.   2.5 in.
1 in.

5. V = _____ m³

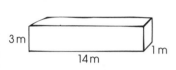

3 m
14 m   1 m

6. V = _____ cm³

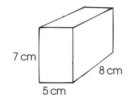

7 cm   8 cm
5 cm

7. V = _____ m³

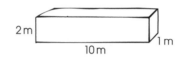

2 m
10 m   1 m

8. V = _____ in.³

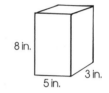

8 in.   3 in.
5 in.

9. V = _____ mm³

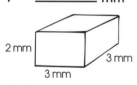

2 mm   3 mm
3 mm

10. V = _____ in.³

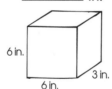

6 in.   3 in.
6 in.

11. V = _____ mm³

3 mm   1 mm
2 mm

12. V = _____ m³

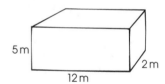

5 m   2 m
12 m

Given the dimensions, find the volume for each rectangular prism.

13. l = 2 cm
w = 4 cm
h = 3 cm

V = _____

14. l = 5 m
w = 3 m
h = 4 m

V = _____

15. l = 10 in.
w = 3 in.
h = 5 in.

V = _____

16. l = 3.5 ft.
w = 1 ft.
h = 2 ft.

V = _____

17. l = 4 m
w = 2.5 m
h = 6 m

V = _____

18. l = 1 cm
w = 20 cm
h = 10 cm

V = _____

19. l = 2 yd.
w = 3 yd.
h = 7 yd.

V = _____

20. l = 10 cm
w = 8 cm
h = 2 cm

V = _____

21. l = 3.4 m
w = 2 m
h = 5 m

V = _____

22. l = 8 yd.
w = 2 yd.
h = 5 yd.

V = _____

Finding Volume

Find the volume of each figure.

1. V = _____

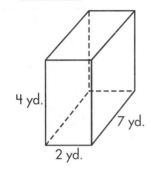

4 yd.
7 yd.
2 yd.

2. V = _____

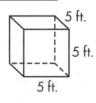

5 ft.
5 ft.
5 ft.

3. V = _____

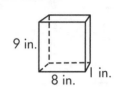

9 in.
8 in.
1 in.

Complete the table to show length, width, height, and volume of each rectangular prism.

	Length	Width	Height	Volume
4.		4 mm	6 mm	48 mm³
5.	8 mm	5 mm	1 mm	
6.	2 mm	9 mm	4 mm	
7.		3 mm	18 mm	162 mm³
8.	12 mm	2 mm		72 mm³
9.		3 mm	3 mm	279 mm³
10.	2 mm	11 mm	4 mm	
11.	6 mm	9 mm		108 mm³
12.	5 mm	4 mm	8 mm	
13.	15 mm		3 mm	270 mm³
14.	12 mm	3 mm	3 mm	
15.	23 mm	3 mm		69 mm³
16.		5 mm	11 mm	550 mm³
17.	9 mm		20 mm	540 mm³
18.	2 mm	14 mm	6 mm	
19.		1 mm	15 mm	60 mm³

Additive Volume

To find the volume of a complex rectangular prism, follow these steps:

1. Break it into rectangular parts.
2. Find the volume of each part.
3. Add the volumes of the parts.

3 cm 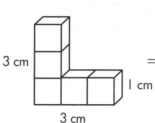 1 cm

= +

3 cm

Find the volume of each figure.

1.

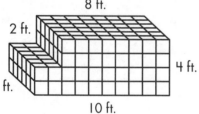

8 ft.
2 ft.
4 ft.
6 ft.
10 ft.

V = _____

2.

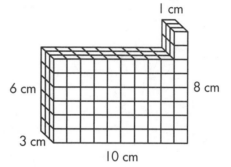

1 cm
6 cm
8 cm
3 cm
10 cm

V = _____

3.

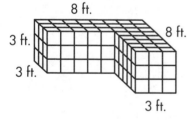

8 ft.
3 ft.
8 ft.
3 ft.
3 ft.

V = _____

4.

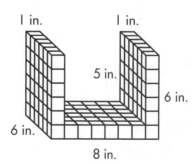

1 in.   1 in.
5 in.
6 in.
6 in.
8 in.

V = _____

Additive Volume

> Remember, to find the volume of a complex rectangular prism, find the volume of each part and then add the volumes of the parts.

Find the volume of each figure.

1.

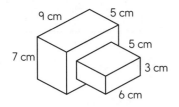

V = _____

2.

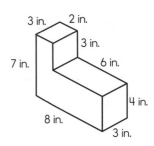

V = _____

3.

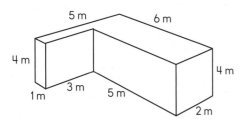

V = _____

4.

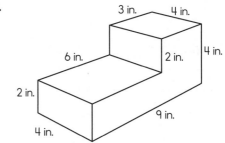

V = _____

5.

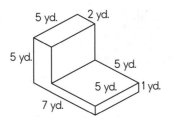

V = _____

6.

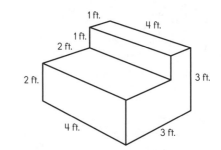

V = _____

Additive Volume

Find the volume of each figure.

1.

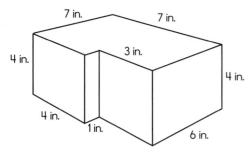

V = _____

2.

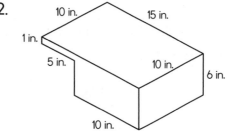

V = _____

3.

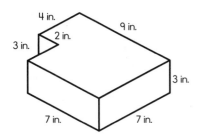

V = _____

4.

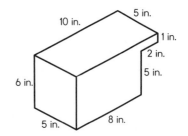

V = _____

5.

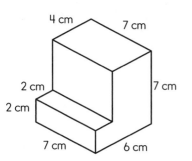

V = _____

6.

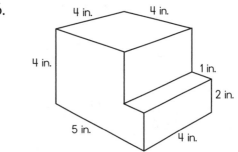

V = _____

7.

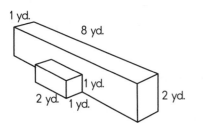

V = _____

8.

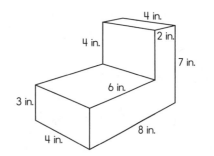

V = _____

Volume Word Problems

Circle the correct answer for each problem.

1. A company measured their cereal box. What is the volume if the dimensions are 2 in. long, 14 in. high, and 2 in. wide?

   A. 18 cubic in.　　　B. 24 cubic in.　　　C. 56 cubic in.　　　D. 48 cubic in.

2. A baby's block measures 12 cm on all sides. What is the volume?

   A. 1,728 cubic cm　　B. 1,200 cubic cm　　C. 144 cubic cm　　　D. 36 cubic cm

3. A juice box measures measures 4 cm long, 10 cm in high, and 5 cm in wide. What is the volume?

   A. 300 cubic cm　　　B. 190 cubic cm　　　C. 19 cubic cm　　　D. 200 cubic cm

4. The dimensions of a toy box are 2 ft. high, 2 ft. wide, and 3 ft. long. What is the volume of the toy box?

   A. 12 cubic ft.　　　B. 7 cubic ft.　　　C. 15 cubic ft.　　　D. 22 cubic ft.

5. A new sandbox measures 12 ft. x 1 ft. x 6 ft. How much room is there for sand?

   A. 19 cubic ft.　　　B. 72 cubic ft.　　　C. 52 cubic ft.　　　D. 22 cubic ft.

Volume Word Problems

Answer each question.

1. Monty and his friend are building a wooden frame for a garden. They want it to measure 10 feet long, 5 feet wide, and 1 foot high. What will the volume of the frame be?

2. The shed in the James's yard is 16 yards long, 4 yards wide, and 5 yards high. How much space is inside the shed?

3. Your neighbors are pouring concrete for their driveway's foundation. The foundation will be 0.25 feet deep, 24 feet long, and 12 feet wide. How much concrete will they need to complete the job?

4. A shipping box is 18 inches long, 12 inches wide, and 6 inches tall. What volume of products can it hold?

5. A bag is 14 inches tall, 10 inches wide, and 4 inches deep. Is it large enough to hold Natalie's school books, which have a combined volume of 500 cubic inches? If so, how much space will she have left? If not, how much more space will she need?

Volume Word Problems

Answer each question.

1. Luke is making a 3-layer square cake. Each layer has a side length of 25 centimeters and a depth of 7 centimeters. What volume of cake batter will he need to make the cake?

2. A student desk is 20 inches wide, 18 inches deep, and 6 inches tall. If each of Mischa's 4 books takes up 144 cubic inches, how much empty space will Mischa have in her desk?

3. Pilar digs a hole in the garden that is 2 feet wide, 4 feet long, and 1 foot deep. Her brother Hector digs a hole that is 4 feet square and 3 feet deep. What volume of dirt did they remove altogether?

4. A ball pit is 6 meters long, 8 meters wide, and 2 meters deep. If about 85 balls fit in 1 cubic meter, about how many balls will it take to fill the ball pit?

5. A swimming pool is 5 yards wide, 9 yards long, and 5 feet deep. If the pool fills at a rate of 100 cubic feet an hour, how long will the pool take to fill?

Graphing Coordinates

Coordinates are like directions for placing a point on a coordinate plane.

(3,4)

- Always start at 0.
- The first number, 3, tells you how many spaces to move over, or along the x-axis.
- The second number, 4, tells you how many spaces to move up, or along the y-axis.

So, for (3,4) you should move over 3 and up 4 to locate the point.

Graph and label each pair of coordinates.

1. A (3,4)

2. B (1,8)

3. C (5,1)

4. D (3,7)

5. E (8,2)

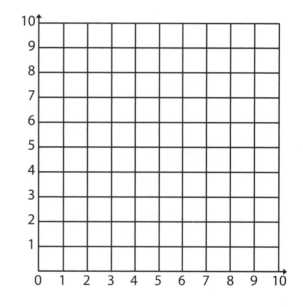

Graph and label each pair of coordinates.

6. F (2,9)

7. G (10,7)

8. H (6,9)

9. I (1,5)

10. J (4,3)

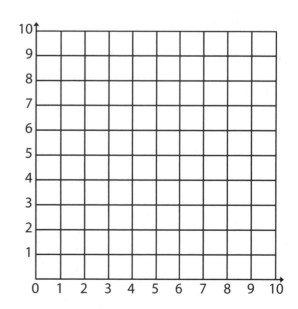

Graphing Coordinates

Remember, in a coordinate pair, the first number tells you how many spaces to move over. The second number tells you how many spaces to move up.

Graph and label each pair of coordinates.

1.   A (3,5)

2.   B (7,8)

3.   C (1,3)

4.   D (6,10)

5.   E (9,4)

6.   F (8,1)

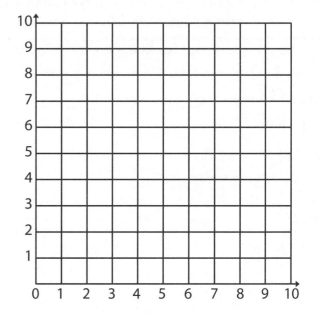

Graph and label each pair of coordinates.

7.   G (2,6)

8.   H (10,3)

9.   I (5,4)

10. J (8,7)

11. K (4,10)

12. L (9,1)

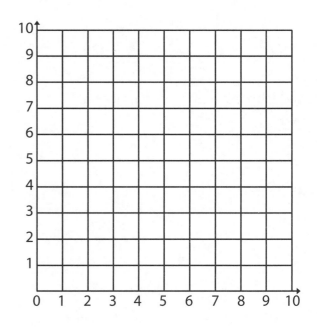

Graphing Coordinates

Graph and label each pair of coordinates.

1. A (6,5)

2. B (1,7)

3. C (10,9)

4. D (8,3)

5. E (5,0)

6. F (7,2)

7. G (0,5)

8. H (2,9)

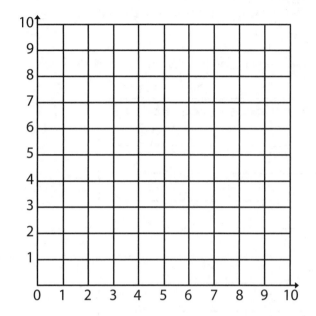

Graph and label each pair of coordinates.

9. I (9,5)

10. J (0,8)

11. K (1,1)

12. L (6,4)

13. M (2,0)

14. N (7,2)

15. O (7,0)

16. P (3,3)

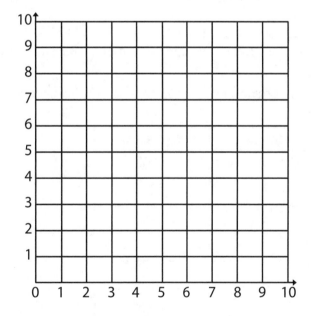

Name _____

Graphing Patterns

Use the patterns to complete the charts. Use the data to plot the information on the graphs. Use the completed graphs to answer the questions.

1. Nina is making necklaces for her friends. Each necklace uses 25 beads.

Necklaces	Number of Beads
1	25
2	50
3	
4	
5	
6	
7	
8	

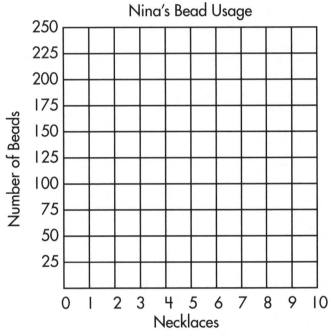

Beads come in packs of 100.
How many necklaces can Nina make with one pack? _____

2. Davis earns $6 each week for doing extra chores at home.

Week	Amount
1	$6
2	$12
3	
4	
5	
6	
7	
8	

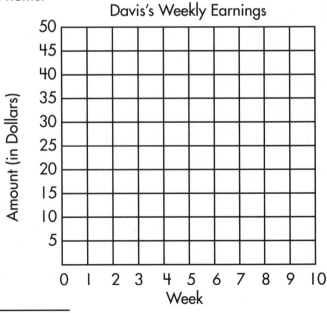

How much money does Davis earn in 6 weeks? _____

Graphing Patterns

Use the patterns to complete the charts. Use the data to plot the information on the graphs. Use the completed graphs to answer the questions.

1. Lily spends 15 minutes cleaning her fish tank every 5 days.

Days	Minutes
5	
10	
15	
20	
25	
30	
35	
40	

After how many days has she spent 2 hours total cleaning the tank? _____

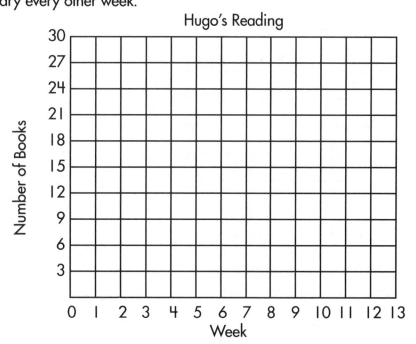

Lily's Cleaning Time

2. Hugo checks out 4 books from the library every other week.

Week	Number of Books
1	
3	
5	
7	
9	
11	
13	

When will Hugo make his goal of reading 25 books?

Hugo's Reading

Graphing Patterns

Use the patterns to complete the charts. Use the data to plot the information on the graphs. Use the completed graphs to answer the questions.

1. Xander spends 20 minutes practicing his guitar every other day.

Day	Time (in Min.)

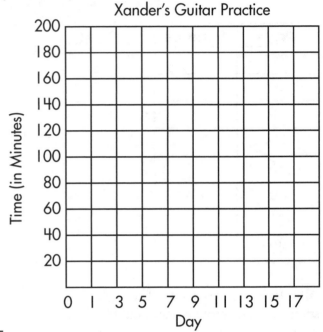

Xander's Guitar Practice

How many days does it take him to practice 3 hours total? _____

2. Ana runs 1.5 laps at soccer practice each week.

Week	Number of Laps

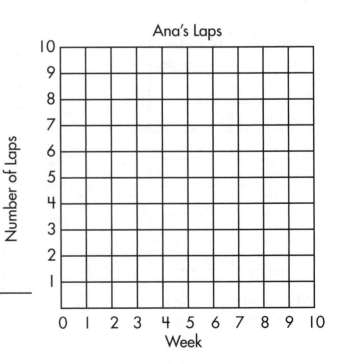

Ana's Laps

How many laps does Ana run every month (every 4 weeks)? _____

Understanding Attributes of Two-Dimensional Figures

Attributes of a shape are the characteristics that define it and separate it from similar shapes. Number of sides and angles, and types of angles and lines are all examples of attributes.

Polygons

equilateral triangle	parallelogram	rectangle	regular hexagon
rhombus	right triangle	square	trapezoid

List the shapes that share each attribute.

1. 4 sides

2. equal sides

3. equal, opposite sides

4. at least one pair of parallel sides

5. 4 angles

6. at least 1 right angle

7. 4 right angles

8. equal, opposite angles

Understanding Attributes of Two-Dimensional Figures

For each shape, write its attributes.

1. rectangle

2. square

3. trapezoid

4. rhombus

5. parallelogram

6. quadrilateral

7. Can a parallelogram be a rectangle? Why or why not?

_____

_____

8. What makes a rhombus different from a square?

_____

_____

9. What attribute makes a trapezoid different from other quadrilaterals?

_____

_____

Name _____

Understanding Attributes of Two-Dimensional Figures

equilateral triangle	parallelogram	rectangle	rhombus
right triangle	square	trapezoid	

Use the polygon names to answer the questions. Each word may be used more than once or not at all.

1. A square is also a _____ because it has 4 right angles and equal, opposite sides.

2. A _____ cannot be a rectangle unless it has 4 right angles.

3. A _____ is only a square if it has 4 right angles.

4. A _____ will never be a rectangle because it only has 1 pair of equal, opposite sides.

5. A _____, _____, and a _____ all have equal sides.

6. An _____ is the only 3-sided figure that can have a right angle.

7. A _____ is the only quadrilateral that can have only 1 right angle.

8. A _____ is like a _____ because both are quadrilaterals with equal and opposite angles that do not have to be right angles.

Answer Key

Name _____ 5.OA.A.1

Evaluating Numerical Expressions

Use the following order to solve and calculate expressions:
1. Solve inside parentheses. $(36 \div 12) \times 2 + 3 \rightarrow (3) \times 2 + 3$
2. Multiply and divide from left to right. $3 \times 2 + 3 \rightarrow 6 + 3$
3. Add and subtract from left to right. $6 + 3 \rightarrow 9$

Solve each expression. Remember to follow the order of operations.

1. $(6 \times 2) + 8 = $ **20**
2. $3 + (8 \times 2) = $ **19**

3. $14 \div 2 + 3 = $ **10**
4. $21 \div 7 \times 2 = $ **6**

5. $(5 \times 2) + 3 = $ **13**
6. $(10 + 10) \div 2 = $ **10**

7. $6 \times (3 + 3) = $ **36**
8. $10 \times 10 \div 25 = $ **4**

9. $(17 - 7) \div 5 = $ **2**
10. $50 \div 5 + 3 = $ **13**

Name _____ 5.OA.A.1

Evaluating Numerical Expressions

Solve each expression. Remember to follow the order of operations.

1. $8 - 2 + 9 = $ **15**
2. $18 \div 9 + 3 = $ **5**

3. $(15 \div 3) \times 5 = $ **25**
4. $(45 \div 9) \times 4 = $ **20**

5. $5 + (21 + 4) = $ **30**
6. $15 - 2 + 8 = $ **21**

7. $40 \div 5 - 7 = $ **1**
8. $(5 \times 7) - 30 = $ **5**

9. $(4 \times 4) + 5 = $ **21**
10. $21 \div 7 + 3 = $ **6**

11. $(2^2 - 2) \times 3 + 8 = $ **14**
12. $(15 \times 2) \div 10 + 8 = $ **11**

13. $10 - (14 \div 2) + 3 = $ **6**
14. $2 \times 2 \div 2 \times 8 = $ **16**

Name _____ 5.OA.A.1

Evaluating Numerical Expressions

Solve each expression.

1. $8 \times (8 - 2) + 6^2 = $ **84**
2. $(52 - 4^2) \div (8 - 4) = $ **9**

3. $(2 + 2)^2 + (14 \div 2) = $ **23**
4. $(12 \times 10 - 4^2) - 5 = $ **99**

5. $(14 - 2)^2 + (8 \div 4) = $ **146**
6. $(64 - 6^2) \div (9 - 2) = $ **4**

7. $6 \times (13 - 4) + 2^2 = $ **58**
8. $[4^2 + (15 \div 5 + 5^2)] + 4^2 = $ **60**

9. $[5^2 + (10 \div 2 + 2^2)] - 5^2 = $ **9**
10. $[(9 - 2)^2 + 3] + 5 + 3^2 = $ **66**

11. $[(10 - 5)^2 \times 7] + 5 + 5^2 = $ **205**
12. $6 + [9 + (9 - 3)^2] + 7 = $ **58**

13. $(16 \div 8)^2 + [(15 - 2) \times 3^2] = $ **121**
14. $13 + [7 + (10 - 6)^2] - 8 = $ **28**

Name _____ 5.OA.A.2

Interpreting Numerical Expressions

Expressions can be written with numbers and symbols or words.

4 more than the product of 6 and 7	add 10 and 12, then divide in half
$4 + (6 \times 7)$	$(10 + 12) \div 2$

Look for keywords to help you decide which operations to use. Use parentheses to group the part of the expression that should happen first.

Write each expression with numbers.

1. 3 times the sum of 2 and 46 $\quad$ **$3 \times (2 + 46)$**
2. 16 more than the product of 2 and 9 $\quad$ **$(2 \times 9) + 16$**
3. subtract 4 from 29, then double $\quad$ **$(29 - 4) \times 2$**
4. 6 less than the quotient of 90 divided by 9 $\quad$ **$(90 \div 9) - 6$**

Write each expression in words.

5. $9 + (24 \div 6)$ **9 more than 24 divided by 6**
6. $(86 - 72) + 6$ **6 more than the difference between 86 and 72**
7. $(22 \times 3) \div 2$ **half of the product of 22 and 3**
8. $4 \times (5 + 83)$ **4 times the sum of 5 and 83**

Answer Key

Name _____   5.OA.A.2

Interpreting Numerical Expressions

Write each expression with numbers.

1. add 34 to itself, then divide in half **(34 + 34) ÷ 2**

2. subtract 6 from 34, then triple **(34 − 6) × 3**

3. 5 times the sum of 23 and 7 **5 × (23 + 7)**

4. the quotient of 100 and double 25 **100 ÷ (25 × 2)**

5. 8 more than the product of 12 and 4 **(12 × 4) + 8**

6. 16 less than the quotient of 200 divided by 10 **(200 ÷ 10) − 16**

Write each expression in words.

7. (33 + 6) ÷ 3 **a third of the sum of 33 and 6**

8. 2 × (15 + 67) **double the sum of 15 and 67**

9. (3 × 13) + 2 **2 more than the product of 3 and 13**

10. (20 − 4) + 100 **100 more than the difference between 20 and 4**

11. (100 − 45) × 2 **double the difference between 100 and 45**

12. 70 + (45 ÷ 9) **70 more than 45 divided by 9**

© Carson-Dellosa • CD-104630   9

Name _____   5.OA.A.2

Interpreting Numerical Expressions

Write each expression with numbers.

1. half of the product of 42 and 3 **(42 × 3) ÷ 2**

2. four times the sum of 53 and 11 **(53 + 11) × 4**

3. the product of 7 and the sum of 5 and 74 **7 × (5 × 74)**

4. 23 times 10, divided by 5 **(23 × 10) ÷ 5**

5. 18 more than the product of 2 and 42 **(2 × 42) + 18**

6. 14 less than the product of 16 and 10 **(16 × 10) − 1**

7. the sum of 45 and the product of 3 and 12 **45 + (3 × 12)**

Write each expression in words.

8. (13 × 6) ÷ 2 **half of the product of 13 and 6**

9. 12 + (30 + 7) **12 plus the sum of 30 and 7**

10. (40 ÷ 8) − 12 **12 less than the quotient of 40 and 8**

11. (21 + 9) ÷ 3 **a third of the sum of 21 and 9**

12. (10 + 2) − (45 ÷ 9) **the difference between the sum of 10 and 2 and the quotient of 49 divided by 9**

13. 80 − (4 × 4) **the difference between 80 and the product of 4 times itself**

10   © Carson-Dellosa • CD-104630

Name _____   5.NBT.A.1, 3.NBT.A.2

Powers of Ten

> Numbers can be abbreviated using exponential notation.
> An exponent tells how many times a factor is multiplied by itself.
> $10^3 = 10 \times 10 \times 10 = 1,000$ 10 is multiplied by itself 3 times.
> Look for patterns when a power of 10 is multiplied by another number between 1 and 9.
> $7,000,000 = 7 \times 10^6$
> Hint: To know what power of 10 to use, simply match the power of 10 to the number of zeros in the number.
> $4,000 = 4 \times 10^3$ $900,000 = 9 \times 10^5$

Write each number with an exponent.

1. 10 to the fourth power **10^4** 2. 10 to the third power **10^3** 3. 10 to the eighth power **10^8**

Solve.

4. 10^3 **1,000** 10^6 **1,000,000** 10^4 **10,000**

5. 10^2 **100** 10^{10} **10,000,000,000** 10^8 **100,000,000**

6. 10^7 **10,000,000** 10^5 **100,000** 10^9 **1,000,000,000**

Rewrite each problem without the exponent. Then, solve.

7. 3×10^2 **= 3 × 100** **= 300** 8×10^3 **= 8 × 1,000** **= 8,000**

8. 6×10^4 **= 6 × 10,000** **= 60,000** 4×10^5 **= 4 × 100,000** **= 400,000**

Write each number as a number multiplied by a power of 10.

9. 7,000 **7×10^3**   5,000 = **5×10^3**   600,000 = **6×10^5**

10. 8,000,000 **8×10^6**   40,000 = **4×10^4**   3,000,000,000 = **3×10^9**

© Carson-Dellosa • CD-104630   11

Name _____   5.NBT.A.1, 3.NBT.A.2

Powers of Ten

Choose numbers to complete each problem. Solve. Look for patterns. **Answers will vary.**

1. ___ × 10 = ___   2. ___ × 100 = ___   3. ___ × 1,000 = ___

___ × 10 = ___   ___ × 100 = ___   ___ × 1,000 = ___

___ × 10 = ___   ___ × 100 = ___   ___ × 1,000 = ___

___ ÷ 10 = ___   ___ ÷ 100 = ___   ___ ÷ 1,000 = ___

___ ÷ 10 = ___   ___ ÷ 100 = ___   ___ ÷ 1,000 = ___

___ ÷ 10 = ___   ___ ÷ 100 = ___   ___ ÷ 1,000 = ___

4. ___ × 0.1 = ___   5. ___ × 0.01 = ___   6. ___ × 0.001 = ___

___ × 0.1 = ___   ___ × 0.01 = ___   ___ × 0.001 = ___

___ × 0.1 = ___   ___ × 0.01 = ___   ___ × 0.001 = ___

___ ÷ 0.1 = ___   ___ ÷ 0.01 = ___   ___ ÷ 0.001 = ___

___ ÷ 0.1 = ___   ___ ÷ 0.01 = ___   ___ ÷ 0.001 = ___

___ ÷ 0.1 = ___   ___ ÷ 0.01 = ___   ___ ÷ 0.001 = ___

7. Write a rule for multiplying and dividing by powers of 10.

Answers will vary but should include that when multiplying, the number of zeros is retained in products. When dividing, the decimal moves left the same amount as the number of zeros in the divisor.

12   © Carson-Dellosa • CD-104630

Answer Key

Name _____ 5.NBT.A.1, 3.NBT.A.2

Powers of Ten

Use what you know about multiplying and dividing by powers of 10 to answer each problem without multiplying or dividing.

1. $6 \times 10,000 =$ **60,000**

   $6 \times 1,000 =$ **6,000**

   $6 \times 100 =$ **600**

   $6 \times 10 =$ **60**

   $6 \times 1 =$ **6**

   $6 \div 1 =$ **6**

   $6 \div 10 =$ **0.6**

   $6 \div 100 =$ **0.06**

   $6 \div 1,000 =$ **0.006**

2. $8 \times 10,000 =$ **80,000**

   $8 \times 1,000 =$ **80,000**

   $8 \times 100 =$ **800**

   $8 \times 10 =$ **80**

   $8 \times 1 =$ **8**

   $8 \div 1 =$ **8**

   $8 \div 10 =$ **0.8**

   $8 \div 100 =$ **0.08**

   $8 \div 1,000 =$ **0.008**

3. $24 \times 10,000 =$ **240,000**

   $24 \times 1,000 =$ **24,000**

   $24 \times 100 =$ **2,400**

   $24 \times 10 =$ **240**

   $24 \times 1 =$ **24**

   $24 \div 1 =$ **24**

   $24 \div 10 =$ **2.4**

   $24 \div 100 =$ **0.24**

   $24 \div 1,000 =$ **0.024**

4. $13 \times 10,000 =$ **130,000**

   $13 \times 1,000 =$ **13,000**

   $13 \times 100 =$ **1,300**

   $13 \times 10 =$ **130**

   $13 \times 1 =$ **13**

   $13 \div 1 =$ **13**

   $13 \div 10 =$ **1.3**

   $13 \div 100 =$ **0.13**

   $13 \div 1,000 =$ **0.013**

© Carson-Dellosa • CD-104630   13

Name _____ 5.NBT.A.3

Understanding Decimals

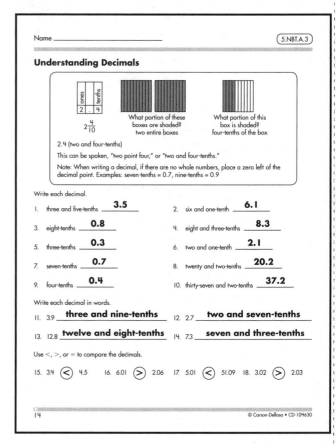

$2\frac{4}{10}$

What portion of these boxes are shaded? two entire boxes

What portion of this box is shaded? four-tenths of the box

2.4 (two and four-tenths)

This can be spoken, "two point four," or "two and four-tenths."

Note: When writing a decimal, if there are no whole numbers, place a zero left of the decimal point. Examples: seven-tenths = 0.7, nine-tenths = 0.9

Write each decimal.

1. three and five-tenths **3.5**

2. six and one-tenth **6.1**

3. eight-tenths **0.8**

4. eight and three-tenths **8.3**

5. three-tenths **0.3**

6. two and one-tenth **2.1**

7. seven-tenths **0.7**

8. twenty and two-tenths **20.2**

9. four-tenths **0.4**

10. thirty-seven and two-tenths **37.2**

Write each decimal in words.

11. 3.9 **three and nine-tenths**

12. 2.7 **two and seven-tenths**

13. 12.8 **twelve and eight-tenths**

14. 7.3 **seven and three-tenths**

Use <, >, or = to compare the decimals.

15. 3.4 **<** 4.5

16. 6.01 **>** 2.06

17. 5.01 **<** 51.09

18. 3.02 **>** 2.03

14   © Carson-Dellosa • CD-104630

Name _____ 5.NBT.A.3

Understanding Decimals

Write each decimal.

1. one-tenth **0.1**

2. twenty-seven hundredths **0.27**

3. three-thousandths **0.003**

4. seven-tenths **0.7**

5. forty-five hundredths **0.45**

6. fifty-one thousandths **0.051**

7. four hundred and one-tenth **400.1**

8. fifty-five and three-tenths **55.3**

9. six-tenths **0.6**

10. one-hundredth **0.01**

Write each decimal in words.

11. 0.04 **four hundredths**

12. 0.99 **ninety-nine hundredths**

13. 0.8 **eight-tenths**

14. 4.89 **four and eighty-nine hundredths**

15. 0.06 **six-hundredths**

Order the numbers in each series from least to greatest.

16. 1.87, 0.187, 10.87

    0.18, 1.87, 10.87

17. 0.045, 0.45, 0.04

    0.04, 0.045, 0.45

18. 0.0065, 0.06, 0.006

    0.06, 0.6, 0.65

19. 0.91, 0.44, 0.23

    0.23, 0.44, 0.91

20. 6.07, 6.17, 6.37

    6.07, 6.17, 6.37

21. 8.98, 8.89, 8.9

    8.89, 8.9, 8.98

© Carson-Dellosa • CD-104630   15

Name _____ 5.NBT.A.3

Understanding Decimals

Write each decimal.

1. three hundred and seven-hundredths **300.07**

2. fifteen and forty-five thousandths **15.045**

3. two hundred eighteen and four-thousandths **218.004**

4. two-thousandths **0.002**

5. sixty-seven and six hundred thirty-one thousandths **67.631**

6. twelve-thousandths **0.012**

7. forty-nine and ninety-nine thousandths **49.099**

8. five and eight hundred forty-five thousandths **5.845**

9. eight-thousandths **0.008**

10. ten and six hundred two-thousandths **10.602**

Write each decimal in words.

11. 0.035 **thirty-five thousandths**

12. 89.004 **eighty-nine and four-thousandths**

13. 324.008 **three hundred twenty-four and eight-thousandths**

14. 72.045 **seventy-two and forty-five thousandths**

Use <, >, or = to compare the decimals.

15. 0.567 **>** 0.423

16. 56.001 **<** 56.01

17. 3.003 **<** 33.003

18. 5.9 **>** 5.09

19. 0.987 **>** 0.789

20. 1.456 **<** 1.665

21. 2.076 **<** 8.076

22. 2.798 **<** 3.009

16   © Carson-Dellosa • CD-104630

Answer Key

Name _____ 5.NBT.A.4

Rounding Decimals

To round a decimal, follow these steps:

1. Underline the place value you are rounding to.

2. If the number to the right of the underline is 0, 1, 2, 3, or 4, the underlined digit stays the same. All the digits to the right change to zeros.

3. If the number to the right of the underline is 5, 6, 7, 8, or 9, the underlined digit goes up by one. All the digits to the right change to zeros.

Examples:
Round to the nearest whole number: 4.8 rounds up to 5.0
Round to the nearest tenth: 14.24 rounds down to 14.20

Round to the nearest whole number.

1. 3.67 **4**   2. 6.8 **7**   3. 11.4 **11**   4. 5.9 **6**

5. 21.24 **21**   6. 10.51 **11**   7. 4.9 **5**   8. 14.2 **14**

9. 8.6 **9**   10. 7.8 **8**   11. 9.21 **9**   12. 10.9 **11**

13. 9.7 **10**   14. 10.3 **10**   15. 8.3 **8**   16. 7.4 **7**

Round to the nearest tenth.

17. 6.29 **6.3**   18. 10.68 **10.7**   19. 14.83 **14.8**   20. 6.84 **6.8**

21. 3.48 **3.5**   22. 24.37 **24.4**   23. 17.47 **17.5**   24. 28.15 **28.2**

25. 5.49 **5.5**   26. 10.43 **10.4**   27. 3.56 **3.6**   28. 6.26 **6.3**

29. 17.64 **17.6**   30. 112.26 **112.3**   31. 9.42 **9.4**   32. 400.67 **400.7**

Name _____ 5.NBT.A.4

Rounding Decimals

Remember, to round a decimal, underline the place value you are rounding to. If the digit to the right is 4 or less, round down. If the digit is 5 or greater, round up.

Round to the nearest whole number.

1. 15.4 **15**   2. 3.9 **4**   3. 58.7 **59**   4. 294.2 **294**

5. 27.3 **27**   6. 18.8 **19**   7. 93.4 **93**   8. 914.5 **915**

Round to the nearest tenth.

9. 25.25 **25.3**   10. 8.16 **8.2**   11. 93.12 **93.1**   12. 724.95 **725.0**

13. 289.37 **289.4**   14. 20.83 **20.8**   15. 204.47 **204.5**   16. 38.52 **38.5**

Round to the nearest hundredth.

17. 205.602 **205.60**   18. 9.995 **10.00**   19. 39.194 **39.19**   20. 532.364 **532.36**

21. 34.436 **34.44**   22. 4.783 **4.78**   23. 93.295 **93.30**   24. 59.364 **59.36**

Name _____ 5.NBT.A.4

Rounding Decimals

Round to the nearest whole number.

1. 45.678 **46**   2. 2.68 **3**   3. 612.123 **612**

4. 2,345.5 **2,346**   5. 7.29 **7**   6. 1.398 **1**

7. 87.056 **87**   8. 43.089 **43**   9. 234.768 **235**

10. 567.289 **567**   11. 7.8 **8**   12. 6.29 **6**

13. 89.519 **90**   14. 4.916 **5**   15. 0.897 **1**

Round to the nearest tenth.

16. 4.379 **4.4**   17. 2.819 **2.8**   18. 543.18 **543.2**

19. 56.14 **56.1**   20. 3.157 **3.2**   21. 78.028 **78.0**

22. 0.417 **0.4**   23. 36.192 **36.2**   24. 1.248 **1.2**

25. 34.248 **34.2**   26. 8.77 **8.8**   27. 16.651 **16.7**

28. 0.54 **0.5**   29. 678.456 **678.5**   30. 0.25 **0.3**

Round to the nearest hundredth.

31. 34.248 **34.25**   32. 5.251 **5.25**   33. 6.108 **6.11**

34. 5.213 **5.21**   35. 9.178 **9.18**   36. 23.682 **23.68**

37. 2.454 **2.45**   38. 9.017 **9.02**   39. 7.271 **7.27**

40. 6.319 **6.32**   41. 2,345.124 **2,345.12**   42. 6.237 **6.24**

43. 45.814 **45.81**   44. 38.199 **38.20**   45. 435.458 **435.46**

Name _____ 5.NBT.B.5

Mulitiplying Multi-Digit Numbers

Solve each problem. Regroup when necessary.

	1.	2.	3.	4.	5.	6.
	323 × 5	515 × 4	255 × 4	915 × 2	860 × 2	561 × 9
	1,615	**2,060**	**1,020**	**1,830**	**1,720**	**5,049**

	7.	8.	9.	10.	11.	12.
	109 × 4	812 × 8	503 × 3	827 × 3	122 × 8	523 × 6
	436	**6,496**	**1,509**	**2,481**	**976**	**3,138**

	13.	14.	15.	16.	17.	18.
	5,306 × 3	6,241 × 7	6,384 × 9	4,634 × 2	8,436 × 5	5,691 × 5
	15,918	**43,687**	**57,456**	**9,268**	**42,180**	**28,455**

	19.	20.	21.	22.	23.	24.
	35 × 28	73 × 56	72 × 43	58 × 63	83 × 27	70 × 60
	980	**4,088**	**3,096**	**3,654**	**2,241**	**4,200**

Answer Key

Multiplying Multi-Digit Numbers

Solve each problem. Regroup when necessary.

#	Problem	Answer	#	Problem	Answer	#	Problem	Answer	#	Problem	Answer	#	Problem	Answer	#	Problem	Answer
1.	87 × 5	**435**	2.	72 × 18	**1,296**	3.	425 × 15	**6,375**	4.	303 × 83	**25,149**	5.	187 × 26	**4,862**	6.	93 × 6	**558**
7.	63 × 25	**1,575**	8.	313 × 72	**22,536**	9.	442 × 81	**35,802**	10.	593 × 45	**26,685**	11.	84 × 3	**252**	12.	42 × 8	**336**
13.	81 × 53	**4,293**	14.	872 × 20	**17,440**	15.	351 × 67	**23,517**	16.	52 × 4	**208**	17.	75 × 21	**1,575**	18.	21 × 10	**210**
19.	214 × 87	**18,618**	20.	109 × 15	**1,635**	21.	12 × 9	**108**	22.	16 × 8	**128**	23.	87 × 26	**2,262**	24.	99 × 21	**2,079**

Multiplying Multi-Digit Numbers

Solve each problem. Regroup when necessary.

#	Problem	Answer	#	Problem	Answer	#	Problem	Answer	#	Problem	Answer	#	Problem	Answer
1.	918 × 55	**50,490**	2.	755 × 221	**166,855**	3.	618 × 500	**309,000**	4.	1,242 × 687	**853,254**	5.	437 × 622	**9,614**
6.	832 × 106	**88,192**	7.	391 × 125	**48,875**	8.	3,861 × 392	**1,513,512**	9.	518 × 42	**21,756**	10.	391 × 535	**209,185**
11.	482 × 663	**319,566**	12.	4,369 × 873	**3,814,137**	13.	925 × 54	**49,950**	14.	851 × 462	**393,162**	15.	331 × 528	**174,768**
16.	7,421 × 694	**5,150,174**	17.	622 × 33	**20,526**	18.	795 × 787	**625,665**	19.	435 × 683	**297,105**	20.	5,872 × 515	**3,024,080**

Division with One-Digit Divisors

Solve each problem.

#	Problem	Answer	#	Problem	Answer	#	Problem	Answer	#	Problem	Answer
1.	4)100	**25**	2.	2)132	**66**	3.	3)225	**75**	4.	9)198	**22**
5.	2)902	**451**	6.	7)112	**16**	7.	6)510	**85**	8.	4)216	**54**
9.	6)426	**71**	10.	2)630	**315**	11.	3)138	**46**	12.	9)369	**41**
13.	8)624	**78**	14.	6)396	**66**	15.	8)648	**81**	16.	5)310	**62**
17.	5)425	**85**	18.	7)672	**96**	19.	3)864	**288**	20.	7)966	**138**

Division with One-Digit Divisors

Solve each problem.

#	Problem	Answer	#	Problem	Answer	#	Problem	Answer	#	Problem	Answer
1.	9)1,368	**152**	2.	4)1,228	**307**	3.	8)5,392	**674**	4.	6)1,878	**313**
5.	5)1,395	**279**	6.	7)2,926	**418**	7.	4)1,008	**252**	8.	5)975	**195**
9.	4)2,128	**532**	10.	2)1,224	**612**	11.	6)2,706	**451**	12.	3)2,019	**673**
13.	3)1,008	**336**	14.	8)3,888	**486**	15.	7)1,421	**203**	16.	5)1,125	**225**
17.	2)1,024	**512**	18.	3)1,134	**378**	19.	8)4,960	**620**	20.	9)2,790	**310**

Answer Key

Division with One-Digit Divisors
Solve each problem.

218r1	**188r3**	**119r5**	**109r6**
1. 4)873	2. 5)943	3. 8)957	4. 9)987

130r5	**105r2**	**298r1**	**108r1**
5. 7)915	6. 5)527	7. 2)597	8. 9)973

143r2	**108r5**	**261r1**	**121r2**
9. 4)574	10. 6)653	11. 3)784	12. 4)486

209r2	**150r1**	**127r2**	**215r2**
13. 3)629	14. 2)301	15. 5)637	16. 4)862

366r1	**117r1**	**191r1**	**163r1**
17. 2)733	18. 8)937	19. 3)574	20. 4)653

Division with Two-Digit Divisors
Solve each problem.

16	**12**	**25**	**24**
1. 32)512	2. 52)624	3. 18)450	4. 32)768

16	**20**	**12**	**30**
5. 62)992	6. 41)820	7. 12)144	8. 32)960

39	**22**	**36**	**15**
9. 18)702	10. 39)858	11. 15)540	12. 23)345

17	**11**	**30**	**13**
13. 56)952	14. 47)517	15. 27)810	16. 26)338

14	**22**	**25**	**16**
17. 25)350	18. 45)990	19. 24)600	20. 54)864

Division with Two-Digit Divisors
Solve each problem.

12r3	**9r36**	**10r14**	**11r30**
1. 67)807	2. 37)369	3. 64)654	4. 81)921

12r9	**6r20**	**36r4**	**38r1**
5. 61)741	6. 58)368	7. 18)652	8. 23)875

18r1	**3r22**	**18r15**	**45r10**
9. 13)235	10. 40)142	11. 25)465	12. 11)505

21r11	**14r10**	**6r49**	**17r22**
13. 19)410	14. 32)458	15. 53)367	16. 45)787

12r4	**15r2**	**11r31**	**4r9**
17. 22)268	18. 42)632	19. 56)647	20. 87)357

Division with Two-Digit Divisors
Solve each problem.

29r9	**52r45**	**24r36**	**332r17**
1. 43)1,256	2. 48)2,541	3. 65)1,596	4. 22)7,321

143r7	**219r22**	**55r2**	**58r2**
5. 21)3,010	6. 39)8,563	7. 82)4,512	8. 37)2,148

205r22	**64r8**	**28r3**	**42r8**
9. 30)6,172	10. 78)5,000	11. 77)2,159	12. 85)3,578

148r55	**181r44**	**120r25**	**105r12**
13. 59)8,787	14. 55)9,999	15. 27)3,265	16. 56)5,892

Answer Key

Name _____ 5.NBT.B.7

Adding Decimals
Solve each problem. Regroup when necessary.

1. $14.2 + 12.1 = $ **26.3**
2. $18.7 + 10.5 = $ **29.2**
3. $1.47 + 6.54 = $ **8.01**
4. $12.3 + 15.2 = $ **27.5**
5. $16.6 + 13.8 = $ **30.4**
6. $7.85 + 9.41 = $ **17.26**

7. $18.2 + 16.5 = $ **34.7**
8. $15.2 + 13.0 = $ **28.2**
9. $2.22 + 3.94 = $ **6.16**
10. $22.2 + 13.1 = $ **35.3**
11. $12.0 + 14.9 = $ **26.9**
12. $7.54 + 2.24 = $ **9.78**

13. $47.5 + 32.6 = $ **80.1**
14. $49.4 + 11.1 = $ **60.5**
15. $8.85 + 7.33 = $ **16.18**
16. $54.8 + 13.2 = $ **68.0**
17. $4.58 + 2.31 = $ **6.89**

18. $12.95 + 5.06 = $ **18.01**
19. $13.8 + 6.9 = $ **20.7**

20. $46.02 + 75.67 = $ **121.69**
21. $16.3 + 35.7 = $ **52.0**

22. $3.25 + 3.25 = $ **6.50**
23. $87.01 + 16.53 = $ **103.54**

Name _____ 5.NBT.B.7

Adding Decimals
Solve each problem.

1. $2.4 + 1.7 = $ **4.1**
2. $18.6 + 9.5 = $ **28.1**
3. $0.01 + 0.72 = $ **0.73**
4. $3.2 + 1.4 + 7.8 = $ **12.4**
5. $2.01 + 3.09 + 8.62 = $ **13.72**

6. $8.1 + 9.2 = $ **17.3**
7. $14.3 + 1.9 = $ **16.2**
8. $1.04 + 2.07 = $ **3.11**
9. $86.7 + 5.2 + 8.4 = $ **100.3**
10. $42.65 + 67.23 + 12.12 = $ **122.00**

11. $10.3 + 7.4 = $ **17.7**
12. $24.7 + 32.6 = $ **57.3**
13. $16.52 + 13.63 = $ **30.15**
14. $9.1 + 12.5 + 19.4 = $ **41.0**
15. $492.6 + 382.3 + 225.7 = $ **1,100.6**

16. $1.5 + 1.5 = $ **3.0**
17. $20.5 + 32.3 = $ **52.8**
18. $14.87 + 56.09 = $ **70.96**
19. $40.08 + 60.27 + 50.33 = $ **150.68**
20. $4.08 + 1.38 + 0.06 = $ **5.52**

Name _____ 5.NBT.B.7

Adding Decimals
Solve each problem.

1. $2.34 + 0.02 + 1.65 = $ **4.01**
2. $543.7 + 3.42 + 0.06 = $ **547.18**
3. $72.56 + 12.38 + 0.07 = $ **85.01**
4. $22.87 + 45.7 + 1.26 = $ **69.83**

5. $987.5 + 4.1 + 30.2 = $ **1,021.8**
6. $2.14 + 0.07 + 72.4 = $ **74.61**
7. $1.70 + 23.75 + 0.05 = $ **25.50**
8. $86.15 + 0.07 + 5.72 = $ **91.94**

9. $5.1 + 7.53 + 87.4 = $ **100.03**
10. $0.2 + 1.2 + 0.12 = $ **1.52**
11. $1.45 + 20.03 + 0.17 = $ **21.65**
12. $4.5 + 5.4 + 12.67 = $ **22.57**

13. $42.7 + 0.03 + 1.7 = $ **44.43**
14. $87.5 + 1.2 + 591.35 = $ **680.05**
15. $0.72 + 1.33 + 12 = $ **14.05**
16. $42 + 0.54 + 7.8 = $ **50.34**

Name _____ 5.NBT.B.7

Subtracting Decimals
Solve each problem. Regroup when necessary.

1. $5.6 - 3.2 = $ **2.4**
2. $10.4 - 8.2 = $ **2.2**
3. $8.5 - 3.5 = $ **5.0**
4. $7.8 - 4.5 = $ **3.3**
5. $9.3 - 7.5 = $ **1.8**

6. $86.5 - 2.3 = $ **84.2**
7. $6.3 - 4.1 = $ **2.2**
8. $8.7 - 5.2 = $ **3.5**
9. $9.65 - 4.22 = $ **5.43**
10. $8.6 - 5.2 = $ **3.4**

11. $16.4 - 8.2 = $ **8.2**
12. $75.4 - 3.1 = $ **72.3**
13. $7.6 - 3.2 = $ **4.4**
14. $26.7 - 2.5 = $ **24.2**
15. $16.2 - 4.1 = $ **12.1**

16. $72.5 - 63.7 = $ **8.8**
17. $8.1 - 6.5 = $ **1.6**

Answer Key

Subtracting Decimals
Solve each problem. Regroup when necessary.

1. 326.7 − 42.8 = **283.9**	2. 14.021 − 5.6 = **8.421**	3. 1.58 − 0.75 = **0.83**	4. 16.88 − 9.3 = **7.58**	5. 52.07 − 3.9 = **48.17**
6. 7.57 − 6.85 = **0.72**	7. 8.12 − 6.01 = **2.11**	8. 18.9 − 16.42 = **2.48**	9. 1.97 − 1.68 = **0.29**	10. 14.9 − 3.2 = **11.7**

11. 19.5 − 0.01 = **19.49**   12. 0.51 − 0.32 = **0.19**

13. 42.62 − 10.35 = **32.27**   14. 28.4 − 4.62 = **23.78**

15. 33.45 − 15.4 = **18.05**   16. 18.5 − 9.5 = **9.0**

Subtracting Decimals
Solve each problem.

1. 19.86 − 1.07 = **18.79**	2. 4.52 − 0.4 = **4.12**	3. 6.25 − 3.01 = **3.24**	4. 23.15 − 3.08 = **20.07**
5. 0.7 − 0.506 = **0.194**	6. 20.34 − 0.3 = **20.04**	7. 756.83 − 22.5 = **734.33**	8. 38.7 − 5.21 = **33.49**
9. 1.42 − 1.2 = **0.22**	10. 71.34 − 2.67 = **68.67**	11. 31.1 − 3.05 = **28.05**	12. 0.65 − 0.224 = **0.43**
13. 2.3 − 1.43 = **0.87**	14. 32.45 − 1.2 = **31.25**	15. 81.38 − 2.77 = **78.61**	16. 24.75 − 6.24 = **18.51**

Multiplying Decimals
Solve each problem. Regroup when necessary.

1. 5.2 × 1.8 = **9.36**	2. 10.5 × 6.6 = **69.3**	3. 2.8 × 9.9 = **27.72**	4. 2.2 × 4.4 = **9.68**	5. 0.12 × 3.7 = **0.444**
6. 5.2 × 0.2 = **1.04**	7. 1.3 × 1.0 = **1.3**	8. 7.1 × 0.25 = **1.775**	9. 7.5 × 2.7 = **20.25**	10. 6.4 × 2.5 = **16**
11. 16.2 × 1.1 = **17.82**	12. 2.0 × 2.1 = **4.2**	13. 5.4 × 1.3 = **7.02**	14. 6.6 × 1.5 = **9.9**	15. 0.44 × 0.1 = **0.044**
16. 0.34 × 0.12 = **0.0408**	17. 5.5 × 4.6 = **25.3**	18. 6.1 × 2.5 = **15.25**	19. 5.6 × 7.3 = **40.88**	20. 3.3 × 0.8 = **2.64**

Multiplying Decimals
Solve each problem. Round to the nearest thousandth when necessary.

1. 0.2 × 4 = **0.8**	2. 0.08 × 6 = **0.48**	3. 5.7 × 0.9 = **5.13**	4. 0.14 × 0.27 = **0.038**	5. 0.67 × 5.4 = **3.618**
6. 0.8 × 3 = **2.4**	7. 0.04 × 5 = **0.2**	8. 4.3 × 2 = **8.6**	9. 7.2 × 5.3 = **38.16**	10. 7.1 × 5.5 = **39.05**
11. 0.5 × 0.4 = **0.2**	12. 6 × 0.12 = **0.72**	13. 1.07 × 0.55 = **0.589**	14. 0.09 × 0.06 = **0.005**	15. 0.07 × 0.15 = **0.011**
16. 0.23 × 0.7 = **0.161**	17. 0.5 × 0.25 = **0.125**	18. 2.15 × 0.8 = **1.72**	19. 1.3 × 3.1 = **4.03**	20. 32 × 6.4 = **204.8**

Answer Key

Page 37

Name _____ 5.NBT.B.7

Multiplying Decimals
Solve each problem. Round to the nearest thousandth when necessary.

1. 0.18×1.5
0.27

2. 0.16×100
16

3. 0.08×0.42
0.0336

4. 87.85×63.4
5,569.69

5. 536.7×1.79
960.693

6. 6.42×3.7
23.754

7. 0.48×13.5
6.48

8. 0.65×53.7
34.905

9. 4.06×0.7
2.842

10. 43.6×64.7
2,820.92

11. 80.42×7.86
632.101

12. 0.62×5.97
3.7014

13. 84.4×0.07
5.908

14. 5.11×0.78
3.986

15. 4.35×0.68
2.958

16. 3.17×0.78
2.4726

Page 38

Name _____ 5.NBT.B.7

Dividing Decimals
Solve each problem.

1. $2\overline{)8.44}$ = **4.22**
2. $5\overline{)1.25}$ = **0.25**
3. $14\overline{)7.21}$ = **0.515**

4. $2\overline{)38.6}$ = **19.3**
5. $67\overline{)656.6}$ = **9.8**
6. $52\overline{)166.4}$ = **3.2**

7. $7\overline{)3.92}$ = **0.56**
8. $2\overline{)9.4}$ = **4.7**
9. $24\overline{)17.28}$ = **0.72**

10. $6\overline{)3.6}$ = **0.6**
11. $7\overline{)3.92}$ = **0.56**
12. $46\overline{)234.6}$ = **5.1**

13. $3\overline{)42.3}$ = **14.1**
14. $5\overline{)72.55}$ = **14.51**
15. $10\overline{)16.6}$ = **1.66**

16. $4\overline{)9.6}$ = **2.4**
17. $5\overline{)8.65}$ = **1.73**
18. $67\overline{)274.7}$ = **4.1**

Page 39

Name _____ 5.NBT.B.7

Dividing Decimals
Solve each problem.

1. $9\overline{)2.7}$ = **0.3**
2. $7\overline{)2.1}$ = **0.3**
3. $4\overline{)0.16}$ = **0.04**
4. $0.8\overline{)56}$ = **70**

5. $6\overline{)3.6}$ = **0.6**
6. $8\overline{)0.64}$ = **0.08**
7. $9\overline{)0.27}$ = **0.03**
8. $0.07\overline{)2.1}$ = **30**

9. $3\overline{)2.7}$ = **0.9**
10. $6\overline{)0.30}$ = **0.05**
11. $0.04\overline{)28}$ = **700**
12. $0.9\overline{)5.4}$ = **6**

13. $3\overline{)0.18}$ = **0.06**
14. $2\overline{)0.12}$ = **0.06**
15. $0.9\overline{)72}$ = **80**
16. $0.7\overline{)0.35}$ = **0.5**

17. $4\overline{)2.4}$ = **0.6**
18. $5\overline{)2.5}$ = **0.5**
19. $0.04\overline{)36}$ = **900**
20. $0.9\overline{)6.3}$ = **7**

Page 40

Name _____ 5.NBT.B.7

Dividing Decimals
Solve each problem.

1. $0.8\overline{)64}$ = **80**
2. $0.25\overline{)100}$ = **400**
3. $6.1\overline{)7.93}$ = **1.3**

4. $0.5\overline{)35}$ = **70**
5. $1.2\overline{)48}$ = **40**
6. $5.3\overline{)42.4}$ = **8**

7. $0.4\overline{)64}$ = **160**
8. $0.7\overline{)4.9}$ = **7**
9. $0.19\overline{)15.2}$ = **80**

10. $0.3\overline{)9}$ = **30**
11. $9.6\overline{)82.8}$ = **8.625**
12. $0.17\overline{)3.23}$ = **19**

13. $0.8\overline{)152}$ = **190**
14. $0.3\overline{)0.63}$ = **2.1**
15. $2.1\overline{)13.65}$ = **6.5**

16. $0.12\overline{)360}$ = **3,000**
17. $0.23\overline{)21.85}$ = **95**
18. $7.2\overline{)40.32}$ = **5.6**

Answer Key

Adding Fractions with Unlike Denominators

1. Find the least common denominator (LCD).	2. Find the equivalent fractions.	3. Add the numerators.
$\frac{1}{3}+\frac{1}{4}$	$\frac{1}{3}\times4=\frac{4}{12}$	$\frac{4}{12}$
LCD = 12	$\frac{1}{4}\times3=\frac{3}{12}$	$+\frac{3}{12}$ → $\frac{7}{12}$

Solve each problem. Write each answer in simplest form.

1. $\frac{1}{4} + \frac{3}{5} = \frac{17}{20}$
2. $\frac{1}{2} + \frac{1}{5} = \frac{7}{10}$
3. $\frac{4}{5} + \frac{2}{3} = 1\frac{7}{15}$
4. $\frac{2}{3} + \frac{2}{5} = 1\frac{1}{15}$
5. $\frac{4}{5} + \frac{7}{8} = 1\frac{27}{40}$

6. $\frac{3}{4} + \frac{1}{3} = 1\frac{1}{12}$
7. $\frac{1}{6} + \frac{2}{5} = \frac{17}{30}$
8. $\frac{3}{6} + \frac{3}{4} = 1\frac{1}{4}$
9. $\frac{1}{4} + \frac{2}{3} = \frac{11}{12}$
10. $\frac{5}{8} + \frac{2}{3} = 1\frac{7}{24}$

11. $\frac{2}{5} + \frac{1}{3} = \frac{11}{15}$
12. $\frac{1}{3} + \frac{5}{7} = 1\frac{1}{21}$
13. $\frac{1}{4} + \frac{7}{8} = 1\frac{1}{8}$
14. $\frac{2}{3} + \frac{3}{15} = \frac{13}{15}$
15. $\frac{2}{6} + \frac{1}{3} = \frac{2}{3}$

Adding Fractions with Unlike Denominators

Solve each problem. Write each answer in simplest form.

1. $\frac{7}{8} + \frac{1}{4} = 1\frac{1}{8}$
2. $\frac{1}{3} + \frac{5}{6} = 1\frac{1}{6}$
3. $\frac{1}{12} + \frac{2}{3} = \frac{3}{4}$
4. $\frac{2}{7} + \frac{2}{5} = \frac{17}{35}$
5. $\frac{3}{10} + \frac{4}{5} = 1\frac{1}{10}$

6. $\frac{1}{12} + \frac{3}{4} = \frac{5}{6}$
7. $\frac{2}{5} + \frac{1}{2} = \frac{9}{10}$
8. $\frac{4}{5} + \frac{3}{6} = 1\frac{3}{10}$
9. $\frac{1}{4} + \frac{1}{2} = \frac{3}{4}$
10. $\frac{2}{3} + \frac{4}{9} = 1\frac{1}{9}$

11. $\frac{1}{8} + \frac{5}{9} = \frac{49}{72}$
12. $\frac{2}{7} + \frac{1}{3} = \frac{13}{21}$
13. $\frac{1}{10} + \frac{4}{8} = \frac{3}{5}$
14. $\frac{5}{8} + \frac{1}{2} = 1\frac{1}{8}$
15. $\frac{2}{3} + \frac{1}{6} = \frac{5}{6}$

16. $\frac{4}{8} + \frac{3}{7} = \frac{13}{14}$
17. $\frac{2}{3} + \frac{5}{6} = 1\frac{1}{2}$
18. $\frac{5}{12} + \frac{1}{4} = \frac{2}{3}$
19. $\frac{6}{12} + \frac{5}{6} = 1\frac{1}{3}$
20. $\frac{1}{3} + \frac{3}{4} = 1\frac{1}{12}$

Adding Fractions with Unlike Denominators

Solve each problem. Write each answer in simplest form.

1. $\frac{2}{5} + \frac{1}{2} = \frac{9}{10}$
2. $\frac{2}{3} + \frac{3}{4} = 1\frac{5}{12}$
3. $\frac{6}{12} + \frac{3}{10} = \frac{4}{5}$
4. $\frac{2}{6} + \frac{1}{8} = \frac{11}{24}$
5. $\frac{3}{10} + \frac{1}{3} = \frac{19}{30}$

6. $\frac{3}{12} + \frac{2}{9} = \frac{17}{36}$
7. $\frac{2}{6} + \frac{5}{12} = \frac{3}{4}$
8. $\frac{4}{8} + \frac{3}{5} = 1\frac{1}{10}$
9. $\frac{7}{8} + \frac{1}{3} = 1\frac{5}{24}$
10. $\frac{5}{6} + \frac{2}{5} = 1\frac{7}{30}$

11. $\frac{1}{7} + \frac{5}{8} = \frac{43}{56}$
12. $\frac{2}{9} + \frac{2}{3} = \frac{8}{9}$
13. $\frac{2}{10} + \frac{3}{4} = \frac{19}{20}$
14. $\frac{5}{6} + \frac{1}{4} = 1\frac{1}{12}$
15. $\frac{1}{12} + \frac{2}{5} = \frac{29}{60}$

16. $\frac{2}{4} + \frac{3}{7} = \frac{13}{14}$
17. $\frac{12}{14} + \frac{4}{5} = 1\frac{23}{35}$
18. $\frac{3}{12} + \frac{2}{14} = \frac{33}{84}$
19. $\frac{5}{13} + \frac{2}{4} = \frac{23}{26}$
20. $\frac{1}{13} + \frac{3}{6} = \frac{15}{26}$

Adding Mixed Numbers with Unlike Denominators

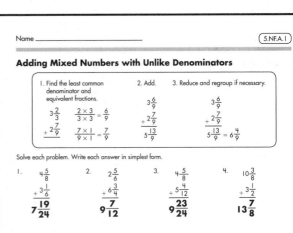

1. Find the least common denominator and equivalent fractions.	2. Add.	3. Reduce and regroup if necessary.
$3\frac{2}{3} \quad \frac{2\times3}{3\times3}=\frac{6}{9}$	$3\frac{6}{9}$	$3\frac{6}{9}$
$+2\frac{7}{9} \quad \frac{7\times1}{9\times1}=\frac{7}{9}$	$+2\frac{7}{9}$	$+2\frac{7}{9}$
	$5\frac{13}{9}$	$5\frac{13}{9}=6\frac{4}{9}$

Solve each problem. Write each answer in simplest form.

1. $4\frac{5}{8} + 3\frac{1}{6} = 7\frac{19}{24}$
2. $2\frac{5}{6} + 6\frac{9}{12} = 9\frac{7}{12}$
3. $4\frac{5}{8} + 5\frac{4}{12} = 9\frac{23}{24}$
4. $10\frac{3}{8} + 3\frac{1}{2} = 13\frac{7}{8}$

5. $3\frac{2}{5} + 2\frac{1}{2} = 5\frac{9}{10}$
6. $8\frac{5}{7} + 9\frac{2}{3} = 18\frac{8}{21}$
7. $8\frac{2}{3} + 1\frac{5}{9} = 10\frac{2}{9}$
8. $2\frac{3}{4} + 7\frac{1}{2} = 10\frac{1}{4}$

9. $1\frac{7}{9} + 4\frac{1}{5} = 5\frac{44}{45}$
10. $6\frac{5}{6} + 2\frac{2}{3} = 9\frac{1}{2}$
11. $4\frac{2}{14} + 6\frac{3}{7} = 10\frac{4}{7}$
12. $1\frac{1}{4} + 5\frac{10}{12} = 7\frac{1}{12}$

Answer Key

Adding Mixed Numbers with Unlike Denominators

Solve each problem. Write each answer in simplest form.

1. $4\frac{1}{8} + 5\frac{3}{4} = 9\frac{7}{8}$
2. $4\frac{7}{8} + 6\frac{1}{4} = 11\frac{1}{8}$
3. $4\frac{3}{4} + 1\frac{2}{3} = 6\frac{5}{12}$

4. $4\frac{1}{8} + 5\frac{1}{5} = 9\frac{13}{40}$
5. $8\frac{3}{4} + 7\frac{3}{16} = 15\frac{15}{16}$
6. $6\frac{1}{2} + 6\frac{2}{5} = 12\frac{9}{10}$

7. $8\frac{1}{3} + 2\frac{3}{7} = 10\frac{16}{21}$
8. $5\frac{1}{8} + 6\frac{2}{5} = 11\frac{21}{40}$
9. $1\frac{9}{10} + 3\frac{1}{4} = 5\frac{3}{20}$

10. $2\frac{3}{4} + 3\frac{5}{6} = 6\frac{7}{12}$
11. $3\frac{1}{9} + 2\frac{1}{3} = 5\frac{4}{9}$
12. $5\frac{2}{3} + 7\frac{3}{7} = 13\frac{2}{21}$

13. $2\frac{5}{6} + 3\frac{1}{3} = 6\frac{1}{6}$
14. $5\frac{1}{2} + 6\frac{2}{7} = 11\frac{11}{14}$
15. $5\frac{5}{15} + 2\frac{3}{5} = 7\frac{3}{5}$

Adding Mixed Numbers with Unlike Denominators

Solve each problem. Write each answer in simplest form.

1. $1\frac{1}{4} + 2\frac{5}{6} = 4\frac{1}{12}$
2. $6\frac{7}{12} + 6\frac{7}{13} = 13\frac{19}{156}$
3. $5\frac{10}{11} + 6\frac{3}{22} = 12\frac{1}{22}$
4. $10\frac{4}{5} + 7\frac{1}{8} = 17\frac{37}{40}$
5. $3\frac{3}{4} + 6\frac{1}{3} = 10\frac{1}{12}$

6. $9\frac{3}{5} + 6\frac{2}{3} = 16\frac{4}{15}$
7. $5\frac{1}{2} + 6\frac{1}{5} = 11\frac{7}{10}$
8. $3\frac{2}{3} + 4\frac{4}{5} = 8\frac{7}{15}$
9. $8\frac{11}{17} + 6\frac{2}{3} = 15\frac{16}{51}$
10. $2\frac{1}{15} + 6\frac{1}{14} = 8\frac{29}{210}$

11. $8\frac{5}{6} + 3\frac{4}{7} = 12\frac{17}{42}$
12. $5\frac{1}{3} + 2\frac{3}{4} = 8\frac{1}{12}$
13. $4\frac{1}{8} + 3\frac{1}{12} = 7\frac{5}{24}$
14. $8\frac{9}{10} + 4\frac{1}{4} = 13\frac{3}{20}$
15. $8\frac{1}{16} + 3\frac{3}{4} = 11\frac{13}{16}$

Subtracting Fractions with Unlike Denominators

1. Find the least common denominator (LCD).	2. Find the equivalent fractions.	3. Subtract. Reduce to simplest form if necessary.
$\frac{3}{4} - \frac{2}{5}$ LCD = 20	$\frac{3}{4} \times 5 = \frac{15}{20}$ $\frac{2}{5} \times 4 = \frac{8}{20}$	$\frac{15}{20} - \frac{8}{20} = \frac{7}{20}$

Solve each problem. Write each answer in simplest form.

1. $\frac{1}{3} - \frac{1}{4} = \frac{1}{12}$
2. $\frac{3}{4} - \frac{1}{5} = \frac{11}{20}$
3. $\frac{9}{10} - \frac{5}{7} = \frac{13}{70}$
4. $\frac{5}{7} - \frac{2}{9} = \frac{31}{63}$
5. $\frac{3}{5} - \frac{1}{3} = \frac{4}{15}$

6. $\frac{3}{8} - \frac{2}{6} = \frac{1}{24}$
7. $\frac{2}{4} - \frac{1}{3} = \frac{1}{6}$
8. $\frac{1}{5} - \frac{1}{8} = \frac{3}{40}$
9. $\frac{7}{12} - \frac{1}{4} = \frac{1}{3}$
10. $\frac{3}{9} - \frac{1}{4} = \frac{1}{12}$

11. $\frac{7}{8} - \frac{1}{9} = \frac{55}{72}$
12. $\frac{8}{8} - \frac{4}{6} = \frac{1}{3}$
13. $\frac{2}{3} - \frac{1}{2} = \frac{1}{6}$
14. $\frac{2}{3} - \frac{4}{9} = \frac{2}{9}$
15. $\frac{1}{3} - \frac{1}{6} = \frac{1}{6}$

Subtracting Fractions with Unlike Denominators

Solve each problem. Write each answer in simplest form.

1. $\frac{3}{9} - \frac{1}{4} = \frac{1}{12}$
2. $\frac{3}{10} - \frac{1}{5} = \frac{1}{10}$
3. $\frac{4}{5} - \frac{5}{10} = \frac{3}{10}$
4. $\frac{15}{27} - \frac{4}{9} = \frac{1}{9}$
5. $\frac{2}{3} - \frac{4}{9} = \frac{2}{9}$

6. $\frac{7}{8} - \frac{2}{16} = \frac{3}{4}$
7. $\frac{2}{4} - \frac{1}{12} = \frac{5}{12}$
8. $\frac{10}{15} - \frac{1}{8} = \frac{13}{24}$
9. $\frac{7}{12} - \frac{2}{4} = \frac{1}{12}$
10. $\frac{3}{9} - \frac{1}{3} = 0$

11. $\frac{7}{8} - \frac{1}{2} = \frac{3}{8}$
12. $\frac{8}{18} - \frac{4}{9} = 0$
13. $\frac{12}{20} - \frac{1}{2} = \frac{1}{10}$
14. $\frac{1}{2} - \frac{1}{4} = \frac{1}{4}$
15. $\frac{8}{10} - \frac{1}{6} = \frac{19}{30}$

16. $\frac{8}{9} - \frac{3}{6} = \frac{7}{18}$
17. $\frac{5}{6} - \frac{1}{5} = \frac{19}{30}$
18. $\frac{7}{8} - \frac{3}{10} = \frac{21}{40}$
19. $\frac{9}{12} - \frac{2}{11} = \frac{25}{44}$
20. $\frac{6}{9} - \frac{3}{12} = \frac{3}{4}$

Answer Key

Page 49

Name _____   5.NF.A.1

Subtracting Fractions with Unlike Denominators
Solve each problem. Write each answer in simplest form.

1. $\frac{3}{4} - \frac{1}{6} = \frac{7}{12}$
2. $\frac{13}{15} - \frac{2}{3} = \frac{1}{5}$
3. $\frac{2}{3} - \frac{7}{12} = \frac{1}{12}$
4. $\frac{5}{6} - \frac{1}{3} = \frac{1}{2}$
5. $\frac{5}{6} - \frac{2}{5} = \frac{13}{30}$

6. $\frac{2}{3} - \frac{1}{6} = \frac{1}{2}$
7. $\frac{11}{14} - \frac{1}{2} = \frac{2}{7}$
8. $\frac{7}{12} - \frac{1}{4} = \frac{1}{3}$
9. $\frac{11}{12} - \frac{1}{16} = \frac{41}{48}$
10. $\frac{5}{6} - \frac{3}{7} = \frac{17}{42}$

11. $\frac{7}{8} - \frac{1}{9} = \frac{55}{72}$
12. $\frac{7}{8} - \frac{1}{12} = \frac{19}{24}$
13. $\frac{5}{12} - \frac{1}{13} = \frac{53}{156}$
14. $\frac{7}{8} - \frac{1}{6} = \frac{17}{24}$
15. $\frac{1}{3} - \frac{1}{6} = \frac{1}{6}$

16. $\frac{2}{3} - \frac{4}{9} = \frac{2}{9}$
17. $\frac{3}{5} - \frac{1}{13} = \frac{35}{52}$
18. $\frac{8}{9} - \frac{5}{6} = \frac{1}{18}$
19. $\frac{9}{12} - \frac{2}{11} = \frac{25}{44}$
20. $\frac{5}{6} - \frac{1}{8} = \frac{17}{24}$

Page 50

Name _____   5.NF.A.1

Subtracting Mixed Numbers with Unlike Denominators

> 1. Find the least common denominator and equivalent fractions.
> $$5\frac{1}{8} \quad \frac{1\times 3}{8\times 3} = \frac{3}{24}$$
> $$-2\frac{1}{3} \quad \frac{1\times 8}{3\times 8} = \frac{8}{24}$$
>
> 2. Borrow and regroup. Subtract the fractions.
> $$5\frac{3}{24} \rightarrow 4\frac{27}{24}$$
> $$-2\frac{8}{24} \rightarrow -2\frac{8}{24}$$
> $$\frac{19}{24}$$
>
> 3. Subtract the whole numbers.
> $$4\frac{27}{24}$$
> $$-2\frac{8}{24}$$
> $$2\frac{19}{24}$$
> Reduce to lowest terms if necessary.

Solve each problem. Write each answer in simplest form.

1. $5\frac{1}{6} - 2\frac{3}{4} = 2\frac{5}{12}$
2. $4\frac{7}{10} - 1\frac{4}{5} = 2\frac{9}{10}$
3. $5\frac{7}{8} - 1\frac{1}{16} = 4\frac{13}{16}$
4. $3\frac{1}{3} - 1\frac{5}{6} = 1\frac{1}{2}$

5. $4\frac{1}{3} - 1\frac{1}{4} = 3\frac{1}{12}$
6. $3\frac{7}{12} - 1\frac{9}{10} = 1\frac{41}{60}$
7. $5\frac{4}{5} - 1\frac{9}{10} = 3\frac{9}{10}$
8. $4\frac{3}{4} - 1\frac{5}{6} = 2\frac{11}{12}$

9. $6\frac{1}{2} - \frac{1}{3} = 6\frac{1}{6}$
10. $7\frac{1}{4} - 3\frac{2}{3} = 3\frac{7}{12}$
11. $10\frac{4}{5} - 6\frac{5}{6} = 3\frac{29}{30}$
12. $12\frac{2}{3} - 9\frac{6}{7} = 2\frac{17}{21}$

Page 51

Name _____   5.NF.A.1

Subtracting Mixed Numbers with Unlike Denominators
Solve each problem. Write each answer in simplest form.

1. $12\frac{7}{8} - 5\frac{5}{16} = 7\frac{9}{16}$
2. $3\frac{1}{4} - 2\frac{5}{12} = \frac{5}{6}$
3. $10\frac{2}{3} - 9\frac{2}{9} = 1\frac{4}{9}$
4. $3\frac{1}{8} - 1\frac{7}{9} = 1\frac{25}{72}$

5. $10\frac{2}{5} - 7\frac{2}{3} = 2\frac{11}{15}$
6. $8\frac{7}{10} - 7\frac{9}{11} = \frac{97}{110}$
7. $8\frac{5}{10} - 7\frac{5}{12} = 1\frac{1}{12}$
8. $5\frac{12}{16} - 5\frac{11}{20} = \frac{1}{5}$

9. $6\frac{1}{6} - 5\frac{5}{12} = \frac{3}{4}$
10. $4\frac{5}{6} - 2\frac{5}{24} = 2\frac{19}{24}$
11. $8\frac{3}{16} - 7\frac{5}{32} = 1\frac{1}{32}$
12. $6\frac{1}{9} - 2\frac{1}{3} = 3\frac{7}{9}$

13. $2\frac{1}{3} - 1\frac{1}{5} = 1\frac{7}{15}$
14. $9\frac{3}{5} - 4\frac{19}{20} = 4\frac{13}{20}$
15. $4\frac{11}{18} - 1\frac{13}{16} = 2\frac{115}{144}$
16. $8\frac{7}{10} - 6\frac{3}{40} = 2\frac{5}{8}$

Page 52

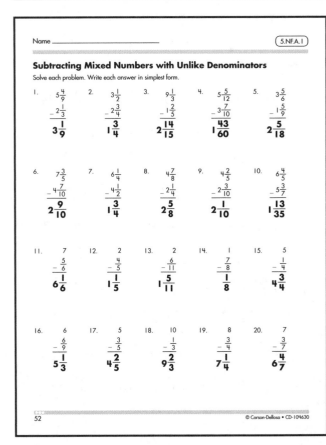

Name _____   5.NF.A.1

Subtracting Mixed Numbers with Unlike Denominators
Solve each problem. Write each answer in simplest form.

1. $5\frac{4}{9} - 2\frac{1}{3} = 3\frac{1}{9}$
2. $3\frac{1}{2} - 1\frac{3}{4} = 1\frac{3}{4}$
3. $9\frac{1}{3} - 6\frac{2}{5} = 2\frac{14}{15}$
4. $5\frac{5}{12} - 3\frac{7}{10} = 1\frac{43}{60}$
5. $3\frac{5}{6} - 1\frac{5}{9} = 2\frac{5}{18}$

6. $7\frac{3}{5} - 4\frac{7}{10} = 2\frac{9}{10}$
7. $6\frac{1}{4} - 4\frac{1}{2} = 1\frac{3}{4}$
8. $4\frac{7}{8} - 2\frac{1}{4} = 2\frac{5}{8}$
9. $4\frac{2}{5} - 2\frac{3}{10} = 2\frac{1}{10}$
10. $6\frac{4}{5} - 5\frac{3}{7} = 1\frac{13}{35}$

11. $7 - \frac{5}{6} = 6\frac{1}{6}$
12. $2 - \frac{4}{5} = 1\frac{1}{5}$
13. $2 - \frac{6}{11} = 1\frac{5}{11}$
14. $1 - \frac{7}{8} = \frac{1}{8}$
15. $5 - \frac{1}{4} = 4\frac{3}{4}$

16. $6 - \frac{6}{9} = 5\frac{1}{3}$
17. $5 - \frac{3}{5} = 4\frac{2}{5}$
18. $10 - \frac{1}{3} = 9\frac{2}{3}$
19. $8 - \frac{3}{4} = 7\frac{1}{4}$
20. $7 - \frac{3}{7} = 6\frac{4}{7}$

Answer Key

5.NBT.B.6, 5.NF.B.3

Name _____

Understanding Fractions as Division
Write each fraction as a division problem. Solve.

1. $\frac{15}{8} = 8\overline{)15}$ **1r7 or 9$\frac{7}{8}$**   2. $\frac{25}{13} = 13\overline{)25}$ **1r12 or 1$\frac{12}{13}$**

3. $\frac{54}{7} = 7\overline{)54}$ **7r5 or 7$\frac{5}{7}$**   4. $\frac{29}{5} = 5\overline{)29}$ **5r4 or 5$\frac{4}{5}$**

5. $\frac{85}{22} = 22\overline{)85}$ **3r19 or 3$\frac{19}{22}$**   6. $\frac{10}{4} = 4\overline{)10}$ **2r2 or 2$\frac{1}{2}$**

7. $\frac{62}{8} = 8\overline{)62}$ **7r6 or 7$\frac{3}{4}$**   8. $\frac{34}{10} = 10\overline{)34}$ **3r4 or 3$\frac{2}{5}$**

9. $\frac{43}{16} = 16\overline{)43}$ **2r11 or 2$\frac{11}{16}$**   10. $\frac{33}{7} = 7\overline{)33}$ **4r5 or 4$\frac{5}{7}$**

53

5.NBT.B.6, 5.NF.B.3

Name _____

Understanding Fractions as Division
Write each fraction as a division problem. Solve.

1. $\frac{16}{7} = 7\overline{)16}$ **2r2 or 2$\frac{2}{7}$**   2. $\frac{32}{9} = 9\overline{)32}$ **3r5 or 3$\frac{5}{9}$**   3. $\frac{4}{3} = 3\overline{)4}$ **1r1 or 1$\frac{1}{3}$**

4. $\frac{37}{12} = 12\overline{)37}$ **3r1 or 3$\frac{1}{12}$**   5. $\frac{12}{5} = 5\overline{)12}$ **2r2 or 2$\frac{2}{5}$**   6. $\frac{20}{15} = 15\overline{)20}$ **1r5 or 1$\frac{1}{3}$**

7. $\frac{50}{27} = 27\overline{)50}$ **1r23 or 1$\frac{23}{27}$**   8. $\frac{10}{4} = 4\overline{)10}$ **2r2 or 2$\frac{1}{2}$**   9. $\frac{7}{4} = 4\overline{)7}$ **1r3 or 1$\frac{3}{4}$**

10. $\frac{53}{16} = 16\overline{)53}$ **3r5 or 3$\frac{5}{16}$**   11. $\frac{86}{19} = 19\overline{)86}$ **4r10 or 4$\frac{10}{19}$**   12. $\frac{55}{12} = 12\overline{)55}$ **4r7 or 4$\frac{7}{12}$**

13. $\frac{14}{13} = 13\overline{)14}$ **1r1 or 1$\frac{1}{13}$**   14. $\frac{43}{20} = 20\overline{)43}$ **2r3 or 2$\frac{3}{20}$**   15. $\frac{18}{5} = 5\overline{)18}$ **3r3 or 3$\frac{3}{5}$**

54

5.NBT.B.6, 5.NF.B.3

Name _____

Understanding Fractions as Division
Write each fraction as a division problem. Solve.

1. $\frac{6}{4} = 4\overline{)6}$ **1r2 or 1$\frac{1}{2}$**   2. $\frac{21}{12} = 12\overline{)21}$ **1r9 or 1$\frac{3}{4}$**   3. $\frac{9}{4} = 4\overline{)9}$ **2r1 or 2$\frac{1}{4}$**

4. $\frac{5}{11} = 11\overline{)5}$ **0r5 or $\frac{5}{11}$**   5. $\frac{19}{5} = 5\overline{)19}$ **3r4 or 3$\frac{4}{5}$**   6. $\frac{3}{2} = 2\overline{)3}$ **1r1 or 1$\frac{1}{2}$**

7. $\frac{7}{4} = 4\overline{)7}$ **1r3 or 1$\frac{3}{4}$**   8. $\frac{13}{3} = 3\overline{)13}$ **4r1 or 4$\frac{1}{3}$**   9. $\frac{14}{6} = 6\overline{)14}$ **2r2 or 2$\frac{1}{3}$**

10. $\frac{16}{5} = 5\overline{)16}$ **3r1 or 3$\frac{1}{5}$**   11. $\frac{3}{5} = 5\overline{)3}$ **0r3 or $\frac{3}{5}$**   12. $\frac{14}{8} = 8\overline{)14}$ **1r6 or 1$\frac{3}{4}$**

13. $\frac{11}{2} = 2\overline{)11}$ **5r1 or 5$\frac{1}{2}$**   14. $\frac{17}{4} = 4\overline{)17}$ **4r1 or 4$\frac{1}{4}$**   15. $\frac{19}{2} = 2\overline{)19}$ **9r1 or 9$\frac{1}{2}$**

16. $\frac{2}{3} = 3\overline{)2}$ **0r2 or $\frac{2}{3}$**   17. $\frac{8}{3} = 3\overline{)8}$ **2r2 or 2$\frac{2}{3}$**   18. $\frac{1}{6} = 6\overline{)1}$ **0r1 or $\frac{1}{6}$**

19. $\frac{1}{3} = 3\overline{)1}$ **0r1 or $\frac{1}{3}$**   20. $\frac{3}{6} = 6\overline{)3}$ **0r3 or $\frac{1}{2}$**   21. $\frac{14}{9} = 9\overline{)14}$ **1r5 or 1$\frac{5}{9}$**

22. $\frac{21}{8} = 8\overline{)21}$ **2r5 or 2$\frac{5}{8}$**   23. $\frac{4}{7} = 7\overline{)4}$ **0r4 or $\frac{4}{7}$**   24. $\frac{13}{3} = 3\overline{)13}$ **4r1 or 4$\frac{1}{3}$**

25. $\frac{12}{5} = 5\overline{)12}$ **2r2 or 2$\frac{2}{5}$**   26. $\frac{8}{11} = 11\overline{)8}$ **0r8 or $\frac{8}{11}$**   27. $\frac{9}{2} = 2\overline{)9}$ **4r1 or 4$\frac{1}{2}$**

28. $\frac{15}{4} = 4\overline{)15}$ **3r3 or 3$\frac{3}{4}$**   29. $\frac{10}{6} = 6\overline{)10}$ **1r4 or 1$\frac{2}{3}$**   30. $\frac{1}{4} = 4\overline{)1}$ **0r1 or $\frac{1}{4}$**

55

5.NF.B.4

Name _____

Multiplying Fractions

> 1. Multiply the numerators. 2. Multiply the denominators. 3. Simplify when necessary.
>
> $\frac{2}{3} \times \frac{5}{6} = 10$ $\frac{2}{3} \times \frac{5}{6} = \frac{10}{18}$ $\frac{10}{18} \div \frac{2}{2} = \frac{5}{9}$

Solve each problem. Write the answer in simplest form.

1. $\frac{3}{4} \times \frac{2}{5} = \frac{3}{10}$ 2. $\frac{7}{8} \times \frac{1}{6} = \frac{7}{48}$ 3. $\frac{4}{5} \times \frac{2}{3} = \frac{8}{15}$

4. $\frac{1}{3} \times \frac{1}{5} = \frac{1}{15}$ 5. $\frac{2}{7} \times \frac{2}{9} = \frac{4}{63}$ 6. $\frac{1}{4} \times \frac{3}{5} = \frac{3}{20}$

7. $\frac{4}{7} \times \frac{3}{8} = \frac{3}{14}$ 8. $\frac{2}{3} \times \frac{2}{5} = \frac{4}{15}$ 9. $\frac{1}{3} \times \frac{3}{5} = \frac{1}{5}$

10. $\frac{3}{5} \times \frac{1}{3} = \frac{1}{5}$ 11. $\frac{1}{8} \times \frac{2}{5} = \frac{1}{20}$ 12. $\frac{1}{6} \times \frac{2}{3} = \frac{1}{9}$

56

Answer Key

Name _____ 5.NF.B.4

Multiplying Fractions
Solve each problem. Write the answer in simplest form.

1. $\frac{1}{3} \times \frac{1}{7} = \mathbf{\frac{1}{21}}$ 2. $\frac{3}{5} \times \frac{2}{9} = \mathbf{\frac{2}{15}}$ 3. $\frac{1}{6} \times \frac{4}{5} = \mathbf{\frac{2}{15}}$

4. $\frac{2}{7} \times \frac{5}{8} = \mathbf{\frac{5}{28}}$ 5. $\frac{2}{5} \times \frac{4}{9} = \mathbf{\frac{8}{45}}$ 6. $\frac{1}{4} \times \frac{1}{6} = \mathbf{\frac{1}{24}}$

7. $\frac{2}{3} \times \frac{3}{8} = \mathbf{\frac{1}{4}}$ 8. $\frac{3}{4} \times \frac{4}{7} = \mathbf{\frac{3}{7}}$ 9. $\frac{2}{5} \times \frac{5}{6} = \mathbf{\frac{1}{3}}$

10. $\frac{4}{5} \times \frac{2}{3} = \mathbf{\frac{8}{15}}$ 11. $\frac{1}{5} \times \frac{5}{6} = \mathbf{\frac{1}{6}}$ 12. $\frac{1}{2} \times \frac{3}{7} = \mathbf{\frac{3}{14}}$

13. $\frac{2}{5} \times \frac{4}{9} = \mathbf{\frac{8}{45}}$ 14. $\frac{2}{8} \times \frac{3}{3} = \mathbf{\frac{1}{4}}$ 15. $\frac{1}{7} \times \frac{6}{8} = \mathbf{\frac{3}{28}}$

Name _____ 5.NF.B.4

Multiplying Fractions
Solve each problem. Write each answer in simplest form.

1. $\frac{1}{4} \times \frac{2}{5} = \mathbf{\frac{1}{10}}$ 2. $\frac{2}{8} \times \frac{3}{6} = \mathbf{\frac{1}{8}}$ 3. $\frac{1}{6} \times \frac{4}{5} = \mathbf{\frac{2}{15}}$

4. $\frac{1}{3} \times \frac{5}{6} = \mathbf{\frac{5}{18}}$ 5. $\frac{4}{6} \times \frac{5}{7} = \mathbf{\frac{10}{21}}$ 6. $\frac{3}{5} \times \frac{1}{8} = \mathbf{\frac{3}{40}}$

7. $\frac{5}{7} \times \frac{2}{4} = \mathbf{\frac{5}{14}}$ 8. $\frac{3}{4} \times \frac{4}{7} = \mathbf{\frac{3}{7}}$ 9. $\frac{3}{4} \times \frac{5}{15} = \mathbf{\frac{1}{20}}$

10. $\frac{12}{16} \times \frac{4}{5} = \mathbf{\frac{3}{5}}$ 11. $\frac{5}{6} \times \frac{3}{4} = \mathbf{\frac{5}{8}}$ 12. $\frac{3}{5} \times \frac{3}{7} = \mathbf{\frac{9}{35}}$

13. $\frac{2}{5} \times \frac{4}{9} = \mathbf{\frac{8}{45}}$ 14. $\frac{12}{18} \times \frac{3}{13} = \mathbf{\frac{1}{7}}$ 15. $\frac{1}{7} \times \frac{3}{4} = \mathbf{\frac{3}{28}}$

Name _____ 5.NF.B.4

Multiplying Whole Numbers and Fractions

1. Convert the whole number to a fraction.	2. Multiply straight across.	3. If the product is an improper fraction, convert to a mixed number in simplest form.
$7 \times \frac{2}{3} = \frac{7}{1} \times \frac{2}{3}$	$\frac{7}{1} \times \frac{2}{3} = \frac{14}{3}$	$\frac{14}{3} = 4\frac{2}{3}$

Solve each problem. Write each answer in simplest form.

1. $5 \times \frac{2}{5} = \mathbf{2}$ 2. $8 \times \frac{1}{7} = \mathbf{1\frac{1}{7}}$ 3. $6 \times \frac{3}{8} = \mathbf{2\frac{1}{4}}$

4. $4 \times \frac{8}{9} = \mathbf{3\frac{5}{9}}$ 5. $2 \times \frac{3}{7} = \mathbf{\frac{6}{7}}$ 6. $\frac{2}{3} \times 4 = \mathbf{2\frac{2}{3}}$

7. $\frac{1}{9} \times 6 = \mathbf{\frac{2}{3}}$ 8. $\frac{5}{6} \times 4 = \mathbf{3\frac{1}{3}}$ 9. $\frac{4}{6} \times 3 = \mathbf{2}$

10. $\frac{4}{5} \times 6 = \mathbf{4\frac{4}{5}}$ 11. $\frac{3}{4} \times 5 = \mathbf{3\frac{3}{4}}$ 12. $2 \times \frac{4}{5} = \mathbf{1\frac{3}{5}}$

Name _____ 5.NF.B.4

Multiplying Whole Numbers and Fractions
Solve each problem. Write each answer in simplest form.

1. $10 \times \frac{2}{3} = \mathbf{6\frac{2}{3}}$ 2. $4 \times \frac{4}{7} = \mathbf{2\frac{2}{7}}$ 3. $7 \times \frac{10}{11} = \mathbf{6\frac{4}{11}}$

4. $36 \times \frac{2}{288} = \mathbf{\frac{1}{4}}$ 5. $6 \times \frac{4}{8} = \mathbf{3}$ 6. $9 \times \frac{5}{6} = \mathbf{7\frac{1}{2}}$

7. $3 \times \frac{1}{3} = \mathbf{1}$ 8. $30 \times \frac{3}{90} = \mathbf{1}$ 9. $12 \times \frac{1}{36} = \mathbf{\frac{1}{3}}$

10. $5 \times \frac{2}{5} = \mathbf{2}$ 11. $12 \times \frac{7}{8} = \mathbf{10\frac{1}{2}}$ 12. $5 \times \frac{3}{4} = \mathbf{3\frac{3}{4}}$

13. $22 \times \frac{1}{44} = \mathbf{\frac{1}{2}}$ 14. $4 \times \frac{1}{8} = \mathbf{\frac{1}{2}}$ 15. $8 \times \frac{2}{3} = \mathbf{5\frac{1}{3}}$

Answer Key

Worksheet 1 (page 61)

Name _____ 5.NF.B.4

Multiplying Whole Numbers and Fractions
Solve each problem. Write each answer in simplest form.

1. $2 \times 2\frac{1}{3} = $ **$4\frac{2}{3}$** 2. $3 \times 5\frac{1}{5} = $ **$15\frac{3}{5}$** 3. $9 \times 3\frac{2}{3} = $ **33**

4. $8 \times 9\frac{1}{10} = $ **$72\frac{4}{5}$** 5. $4 \times 5\frac{1}{8} = $ **$20\frac{1}{2}$** 6. $6 \times 3\frac{1}{6} = $ **19**

7. $5 \times 6\frac{5}{8} = $ **$33\frac{1}{8}$** 8. $3 \times 9\frac{1}{3} = $ **28** 9. $7 \times 1\frac{3}{4} = $ **$12\frac{1}{4}$**

10. $7 \times 2\frac{3}{5} = $ **$18\frac{1}{5}$** 11. $4 \times 2\frac{1}{2} = $ **10** 12. $7 \times 2\frac{1}{7} = $ **15**

13. $3 \times 1\frac{15}{16} = $ **$5\frac{13}{16}$** 14. $4 \times 8\frac{6}{7} = $ **$35\frac{3}{7}$** 15. $2 \times 2\frac{1}{4} = $ **$4\frac{1}{2}$**

© Carson-Dellosa • CD-104630   61

Worksheet 2 (page 62)

Name _____ 5.NF.B.4

Multiplying Mixed Numbers

1. Convert the mixed numbers to fractions.	2. Multiply.	3. Convert the product back to a mixed number in lowest terms.
$1\frac{1}{3} \times 2\frac{1}{2} = \frac{4}{3} \times \frac{5}{2}$	$\frac{4}{3} \times \frac{5}{2} = \frac{20}{6}$	$\frac{20}{6} = 3\frac{2}{6} = 3\frac{1}{3}$

Solve each problem. Write the answer in simplest form.

1. $8\frac{1}{4} \times 6\frac{2}{3} = $ **55** 2. $7\frac{2}{5} \times 6\frac{2}{3} = $ **$49\frac{1}{3}$**

3. $2\frac{5}{6} \times 12\frac{4}{5} = $ **$36\frac{4}{5}$** 4. $4\frac{2}{7} \times 6\frac{1}{10} = $ **$26\frac{1}{7}$**

5. $5\frac{1}{5} \times 4\frac{1}{3} = $ **$22\frac{8}{15}$** 6. $9\frac{9}{10} \times 4\frac{7}{8} = $ **$48\frac{21}{80}$**

7. $1\frac{10}{13} \times 2\frac{9}{13} = $ **$4\frac{129}{169}$** 8. $8\frac{3}{5} \times 4\frac{5}{6} = $ **$41\frac{17}{30}$**

62   © Carson-Dellosa • CD-104630

Worksheet 3 (page 63)

Name _____ 5.NF.B.4

Multiplying Mixed Numbers
Solve each problem. Write the answer in simplest form.

1. $3\frac{1}{2} \times 2\frac{1}{2} = $ **$8\frac{3}{4}$** 2. $8\frac{5}{6} \times 3\frac{6}{7} = $ **$34\frac{1}{14}$** 3. $4\frac{2}{5} \times 6\frac{2}{3} = $ **$29\frac{1}{3}$**

4. $4\frac{2}{9} \times 5\frac{10}{11} = $ **$29\frac{94}{99}$** 5. $2\frac{2}{3} \times 4\frac{2}{5} = $ **$11\frac{11}{15}$** 6. $5\frac{3}{4} \times 6\frac{1}{4} = $ **$35\frac{15}{16}$**

7. $2\frac{8}{9} \times 7\frac{7}{8} = $ **$22\frac{3}{4}$** 8. $7\frac{1}{4} \times 3\frac{3}{7} = $ **$24\frac{6}{7}$** 9. $6\frac{7}{8} \times 3\frac{1}{3} = $ **$22\frac{11}{12}$**

10. $7\frac{9}{10} \times 8\frac{7}{8} = $ **$70\frac{9}{80}$** 11. $4\frac{1}{4} \times 3\frac{5}{6} = $ **$16\frac{7}{24}$** 12. $8\frac{3}{5} \times 1\frac{1}{2} = $ **$12\frac{9}{10}$**

© Carson-Dellosa • CD-104630   63

Worksheet 4 (page 64)

Name _____ 5.NF.B.4

Multiplying Mixed Numbers
Solve each problem. Write each answer in simplest form.

1. $8\frac{1}{4} \times 6\frac{2}{3} = $ **$57\frac{3}{26}$** 2. $7\frac{2}{5} \times 9\frac{1}{8} = $ **$67\frac{21}{40}$** 3. $12\frac{4}{6} \times 2\frac{4}{15} = $ **$28\frac{32}{45}$**

4. $4\frac{2}{17} \times 6\frac{9}{10} = $ **$24\frac{5}{7}$** 5. $2\frac{9}{10} \times 5\frac{7}{8} = $ **$17\frac{3}{80}$** 6. $5\frac{1}{3} \times 4\frac{1}{2} = $ **24**

7. $3\frac{1}{3} \times 3\frac{1}{3} = $ **$11\frac{1}{9}$** 8. $3\frac{3}{4} \times 2\frac{1}{3} = $ **$8\frac{3}{4}$** 9. $5\frac{10}{15} \times 3\frac{2}{3} = $ **$20\frac{7}{9}$**

10. $9\frac{9}{10} \times 4\frac{7}{8} = $ **$48\frac{21}{80}$** 11. $11\frac{1}{3} \times 4\frac{7}{13} = $ **$51\frac{17}{39}$** 12. $10\frac{8}{15} \times 4\frac{2}{6} = $ **$45\frac{29}{45}$**

64   © Carson-Dellosa • CD-104630

Answer Key

Dividing Whole Numbers by Unit Fractions

$4 \div \frac{1}{4}$

Divide each figure into fourths.

$4 \div \frac{1}{4} = 16$ There are now 16 parts.

1. To solve, turn the division problem into a multiplication problem by flipping the digits in the fraction.

$4 \times \frac{4}{1} = 16$

2. Check your quotient by multiplying it by the divisor.

$16 \times \frac{1}{4} = 4$

$\frac{16}{4} = 4$

Solve each problem. Write the answer in simplest form.

1. $1 \div \frac{1}{4} =$ **4**

2. $3 \div \frac{1}{8} =$ **24**

3. $5 \div \frac{1}{10} =$ **50**

4. $1 \div \frac{1}{7} =$ **7**

5. $2 \div \frac{1}{8} =$ **16**

6. $2 \div \frac{1}{2} =$ **4**

7. $2 \div \frac{1}{8} =$ **16**

8. $4 \div \frac{1}{2} =$ **8**

9. $6 \div \frac{1}{7} =$ **42**

Dividing Whole Numbers by Unit Fractions

Solve each problem. Write each answer in simplest form.

1. $1 \div \frac{1}{4} =$ **4**

2. $2 \div \frac{1}{3} =$ **6**

3. $6 \div \frac{1}{10} =$ **60**

4. $1 \div \frac{1}{8} =$ **8**

5. $3 \div \frac{1}{4} =$ **12**

6. $5 \div \frac{1}{8} =$ **40**

7. $2 \div \frac{1}{6} =$ **12**

8. $2 \div \frac{1}{5} =$ **10**

9. $1 \div \frac{1}{3} =$ **3**

10. $3 \div \frac{1}{10} =$ **30**

11. $1 \div \frac{1}{5} =$ **5**

12. $5 \div \frac{1}{4} =$ **20**

Dividing Whole Numbers by Unit Fractions

Solve each problem. Write each answer in simplest form.

1. $1 \div \frac{1}{2} =$ **2**

2. $6 \div \frac{1}{3} =$ **18**

3. $3 \div \frac{1}{10} =$ **30**

4. $1 \div \frac{1}{5} =$ **5**

5. $11 \div \frac{1}{5} =$ **55**

6. $2 \div \frac{1}{2} =$ **4**

7. $2 \div \frac{1}{3} =$ **6**

8. $4 \div \frac{1}{18} =$ **72**

9. $12 \div \frac{1}{2} =$ **24**

10. $4 \div \frac{1}{5} =$ **20**

11. $3 \div \frac{1}{6} =$ **18**

12. $3 \div \frac{1}{8} =$ **24**

13. $8 \div \frac{1}{12} =$ **96**

14. $5 \div \frac{1}{9} =$ **45**

15. $9 \div \frac{1}{15} =$ **135**

Dividing Unit Fractions by Whole Numbers

1. Change the whole number to a fraction.

$\frac{1}{3} \div 6 = \frac{1}{3} \div \frac{6}{1}$

$\frac{1}{3} \div 6 = \frac{1}{3} \times \frac{1}{6}$

2. Find the reciprocal of the second fraction by flipping it. Change the division sign to a multiplication sign.

$\frac{6}{1} = \frac{1}{6}$

3. Multiply. Simplify when necessary.

$\frac{1}{3} \times \frac{1}{6} = \frac{1}{18}$

Solve each problem. Write each answer in simplest form.

1. $\frac{1}{3} \div 4 = \dfrac{1}{12}$

2. $\frac{1}{2} \div 1 = \dfrac{1}{2}$

3. $\frac{1}{3} \div 1 = \dfrac{1}{3}$

4. $\frac{1}{3} \div 2 = \dfrac{1}{6}$

5. $\frac{1}{4} \div 2 = \dfrac{1}{8}$

6. $\frac{1}{8} \div 1 = \dfrac{1}{8}$

7. $\frac{1}{3} \div 8 = \dfrac{1}{24}$

8. $\frac{1}{5} \div 3 = \dfrac{1}{15}$

9. $\frac{1}{3} \div 3 = \dfrac{1}{9}$

Answer Key

Name _____
5.NF.B.7

Dividing Unit Fractions by Whole Numbers
Solve each problem. Write the answer in simplest form.

1. $\frac{1}{4} \div 5 = \frac{1}{20}$ 2. $\frac{1}{5} \div 8 = \frac{1}{40}$ 3. $\frac{1}{3} \div 5 = \frac{1}{15}$

4. $\frac{1}{7} \div 2 = \frac{1}{14}$ 5. $\frac{1}{5} \div 2 = \frac{1}{10}$ 6. $\frac{1}{5} \div 1 = \frac{1}{5}$

7. $\frac{1}{2} \div 3 = \frac{1}{6}$ 8. $\frac{1}{3} \div 15 = \frac{1}{45}$ 9. $\frac{1}{5} \div 3 = \frac{1}{15}$

10. $\frac{1}{5} \div 6 = \frac{1}{30}$ 11. $\frac{1}{4} \div 2 = \frac{1}{8}$ 12. $\frac{1}{8} \div 3 = \frac{1}{24}$

© Carson-Dellosa • CD-104630
69

Name _____
5.NF.B.7

Dividing Unit Fractions by Whole Numbers
Solve each problem. Write each answer in simplest form.

1. $\frac{1}{3} \div 1 = \frac{1}{3}$ 2. $\frac{1}{2} \div 1 = \frac{1}{2}$ 3. $\frac{1}{6} \div 1 = \frac{1}{6}$

4. $\frac{1}{6} \div 4 = \frac{1}{24}$ 5. $\frac{1}{5} \div 2 = \frac{1}{10}$ 6. $\frac{1}{4} \div 1 = \frac{1}{4}$

7. $\frac{1}{4} \div 11 = \frac{1}{44}$ 8. $\frac{1}{8} \div 4 = \frac{1}{32}$ 9. $\frac{1}{2} \div 3 = \frac{1}{6}$

10. $\frac{1}{3} \div 2 = \frac{1}{6}$ 11. $\frac{1}{5} \div 11 = \frac{1}{55}$ 12. $\frac{1}{5} \div 3 = \frac{1}{15}$

13. $\frac{1}{9} \div 8 = \frac{1}{72}$ 14. $\frac{1}{7} \div 4 = \frac{1}{28}$ 15. $\frac{1}{12} \div 7 = \frac{1}{84}$

70
© Carson-Dellosa • CD-104630

Name _____
5.NF.A.2

Fraction Word Problems
Create a story problem that makes sense with each problem. Then, solve.

1. $\frac{4}{6} + \frac{7}{8}$ **Answers will vary.** $1\frac{13}{24}$

2. $5\frac{6}{9} - \frac{1}{5}$ **Answers will vary.** $5\frac{1}{6}$

3. $4\frac{3}{4} + 1\frac{1}{3}$ **Answers will vary.** $6\frac{1}{12}$

4. $6 - 1\frac{2}{5}$ **Answers will vary.** $5\frac{3}{5}$

© Carson-Dellosa • CD-104630
71

Name _____
5.NF.B.6

Fraction Word Problems
Create a story problem that makes sense with each problem. Then, solve.

1. $\frac{1}{4} \times \frac{3}{5}$ **Answers will vary.** $\frac{3}{20}$

2. $\frac{5}{11} \times \frac{2}{3}$ **Answers will vary.** $\frac{10}{33}$

3. $1\frac{4}{5} \times \frac{4}{5}$ **Answers will vary.** $\frac{36}{25}$

4. $8 \times \frac{5}{8}$ **Answers will vary.** 5

72
© Carson-Dellosa • CD-104630

© Carson-Dellosa • CD-104630

Answer Key

Name _____ 5.NF.B.7c

Fraction Word Problems
Create a story problem that makes sense with each problem. Then, solve.

1. $10 \div \frac{1}{4}$ **Answers will vary. 40**

2. $7 \div \frac{1}{3}$ **Answers will vary. 21**

3. $\frac{1}{6} \div 9$ **Answers will vary. $\frac{1}{54}$**

4. $\frac{1}{5} \div 4$ **Answers will vary. $\frac{1}{20}$**

© Carson-Dellosa • CD-104630   73

Name _____ 5.MD.A.1

Converting Measurements

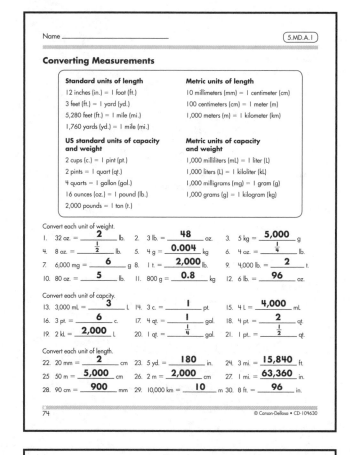

Standard units of length	Metric units of length
12 inches (in.) = 1 foot (ft.)	10 millimeters (mm) = 1 centimeter (cm)
3 feet (ft.) = 1 yard (yd.)	100 centimeters (cm) = 1 meter (m)
5,280 feet (ft.) = 1 mile (mi.)	1,000 meters (m) = 1 kilometer (km)
1,760 yards (yd.) = 1 mile (mi.)	
US standard units of capacity and weight	**Metric units of capacity and weight**
2 cups (c.) = 1 pint (pt.)	1,000 milliliters (mL) = 1 liter (L)
2 pints = 1 quart (qt.)	1,000 liters (L) = 1 kiloliter (kL)
4 quarts = 1 gallon (gal.)	1,000 milligrams (mg) = 1 gram (g)
16 ounces (oz.) = 1 pound (lb.)	1,000 grams (g) = 1 kilogram (kg)
2,000 pounds = 1 ton (t.)	

Convert each unit of weight.
1. 32 oz. = **2** lb. 2. 3 lb. = **48** oz. 3. 5 kg = **5,000** g
4. 8 oz. = **$\frac{1}{2}$** lb. 5. 4 g = **0.004** kg 6. 4 oz. = **$\frac{1}{4}$** lb.
7. 6,000 mg = **6** g 8. 1 t. = **2,000** lb. 9. 4,000 lb. = **2** t.
10. 80 oz. = **5** lb. 11. 800 g = **0.8** kg 12. 6 lb. = **96** oz.

Convert each unit of capcity.
13. 3,000 mL = **3** L 14. 3 c. = **1** pt. 15. 4 L = **4,000** mL
16. 3 pt. = **6** c. 17. 4 qt. = **1** gal. 18. 4 pt. = **2** qt.
19. 2 kL = **2,000** L 20. 1 qt. = **$\frac{1}{4}$** gal. 21. 1 pt. = **$\frac{1}{2}$** qt.

Convert each unit of length.
22. 20 mm = **2** cm 23. 5 yd. = **180** in. 24. 3 mi. = **15,840** ft.
25. 50 m = **5,000** cm 26. 2 m = **2,000** cm 27. 1 mi. = **63,360** in.
28. 90 cm = **900** mm 29. 10,000 km = **10** m 30. 8 ft. = **96** in.

74   © Carson-Dellosa • CD-104630

Name _____ 5.MD.A.1

Converting Measurements
Convert.

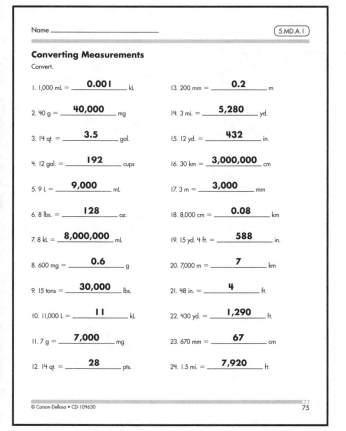

1. 1,000 mL = **0.001** kL
2. 40 g = **40,000** mg
3. 14 qt. = **3.5** gal.
4. 12 gal. = **192** cups
5. 9 L = **9,000** mL
6. 8 lbs. = **128** oz.
7. 8 kL = **8,000,000** mL
8. 600 mg = **0.6** g
9. 15 tons = **30,000** lbs.
10. 11,000 L = **11** kL
11. 7 g = **7,000** mg
12. 14 qt. = **28** pts.

13. 200 mm = **0.2** m
14. 3 mi. = **5,280** yd.
15. 12 yd. = **432** in.
16. 30 km = **3,000,000** cm
17. 3 m = **3,000** mm
18. 8,000 cm = **0.08** km
19. 15 yd. 4 ft. = **588** in.
20. 7,000 m = **7** km
21. 48 in. = **4** ft.
22. 430 yd. = **1,290** ft.
23. 670 mm = **67** cm
24. 1.5 mi. = **7,920** ft.

© Carson-Dellosa • CD-104630   75

Name _____ 5.MD.A.1

Converting Measurements
Convert.

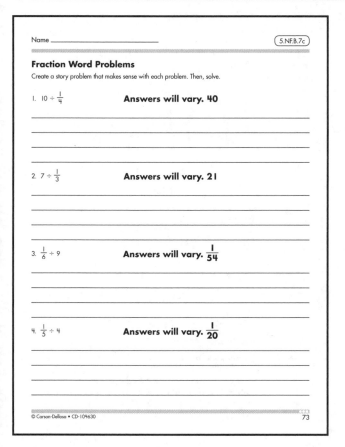

Length	Capacity	Weight
1. 3 ft. = **36** in.	12. 4 c. = **2** pt.	23. 24 oz. = **1.5** lb.
2. 48 in. = **4** ft.	13. 4,600 mL = **4.6** L	24. 4,000 lb. = **2** t.
3. 1,500 cm = **1.5** m	14. 8 qt. = **2** gal.	25. 17,000 g = **17** kg
4. 4 yd. = **12** ft.	15. 1.5 L = **1,500** mL	26. 3 kg 300 g = **3,300** g
5. 15 ft. = **5** yd.	16. 7 pt. = **3.5** qt.	27. 2 lb. = **32** oz.
6. 300 mm = **0.3** m	17. 3 gal. = **12** qt.	28. 36,000 g = **36** kg
7. 4.5 km = **4,500** m	18. 15 kL 200 L = **15,200** L	29. 3.5 t. = **7,000** lb.
8. 2.5 mi. = **4,400** yd.	19. 2 qt. = **8** c.	30. 5 lb. = **80** oz.
9. 6.5 yd. = **234** in.	20. 3 pt. = **6** c.	31. 8.5 kg = **8,500** g
10. 60 in. = **5** ft.	21. 2,000,000 mL = **2** kL	32. 3 lb. = **48** oz.
11. 1.2 mi. = **1,200,000**	22. 4.5 gal. = **18** qt.	33. 10,000 lb. = **5** t.

Use <, >, or = to compare.

34. 8,000 mg **<** 10 g 35. 7 g **>** 60 mg 36. 800 mg **<** 8 g
37. 10 kg **=** 10,000 g 38. 5 g **>** 10 mg 39. 18 mg **<** 20 g
40. 100 mm **=** 10 cm 41. 10 km **>** 20 m 42. 3 cm **<** 35 mm
43. 5 cm **>** 10 mm 44. 1 km **>** 300 m 45. 2 km **=** 2,000 m

76   © Carson-Dellosa • CD-104630

© Carson-Dellosa • CD-104630   121

Answer Key

Panel 1 (page 77):

Name _____   5.MD.B.2

Reading Line Plots

Use the line plot to answer the questions.

Lengths of Sticks in Inches

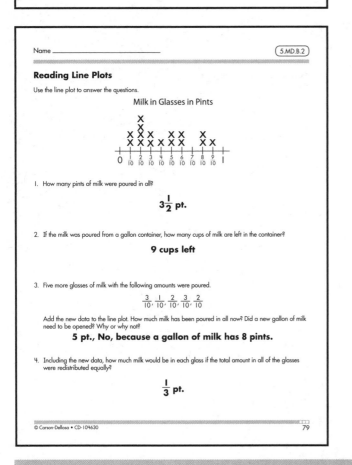

1. How many sticks were part of the data set?

   13 sticks

2. How many sticks were shorter than 8.75 inches?

   10 sticks

3. Maya needs sticks from 8.25 inches to 8.75 inches for a craft project. What fraction of the sticks collected can she use?

   She can use $\frac{10}{13}$ of the sticks.

4. Maya found 8 more sticks, with the following measurements.

   $8\frac{1}{8}, 8\frac{1}{8}, 8, 9, 8\frac{5}{8}, 8\frac{3}{8}, 8\frac{5}{8}, 8\frac{2}{8}$

   Add the new data to the line plot. Now what fraction of the set can Maya use?

   She can use $\frac{2}{3}$ of the sticks.

Panel 2 (page 78):

Name _____   5.MD.B.2

Reading Line Plots

Use the line plot to answer the questions.

Length of String in Inches

1. How many pieces of string were 4 inches or longer?

   7 pieces

2. What fraction of the pieces are not exactly 3, 4, or 5 inches?

   $\frac{7}{10}$ of the pieces

3. If the pieces were all cut from the same piece of string, how long was the original piece of string?

   $40\frac{1}{4}$ in.

4. Five more pieces of string were cut, with the following lengths:

   $3\frac{3}{4}, 3, 4\frac{1}{4}, 4\frac{3}{4}, 3\frac{3}{4}$

   Add the new data to the line plot. Could all of the pieces on the line plot have been cut from a 5-foot piece of string? Why or why not?

   Yes, because they total $59\frac{3}{4}$ inches, which is less than 5 feet, or 60 inches.

Panel 3 (page 79):

Name _____   5.MD.B.2

Reading Line Plots

Use the line plot to answer the questions.

Milk in Glasses in Pints

1. How many pints of milk were poured in all?

   $3\frac{1}{2}$ pt.

2. If the milk was poured from a gallon container, how many cups of milk are left in the container?

   9 cups left

3. Five more glasses of milk with the following amounts were poured.

   $\frac{3}{10}, \frac{1}{10}, \frac{2}{10}, \frac{3}{10}, \frac{2}{10}$

   Add the new data to the line plot. How much milk has been poured in all now? Did a new gallon of milk need to be opened? Why or why not?

   5 pt., No, because a gallon of milk has 8 pints.

4. Including the new data, how much milk would be in each glass if the total amount in all of the glasses were redistributed equally?

   $\frac{1}{3}$ pt.

Panel 4 (page 80):

Name _____   5.MD.C.3, 5.MD.C.4

Exploring Volume

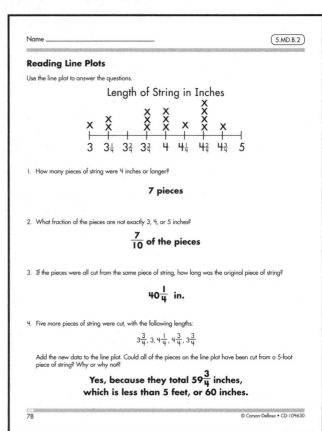

Volume tells the number of cubic units inside a figure. Each box represents one cubic unit.

4 cubic units

Write the number of cubic units in each figure.

1. V = **4** cubic units

2. V = **16** cubic units

3. V = **35** cubic units

4. V = **18** cubic units

5. V = **8** cubic units

6. V = **20** cubic units

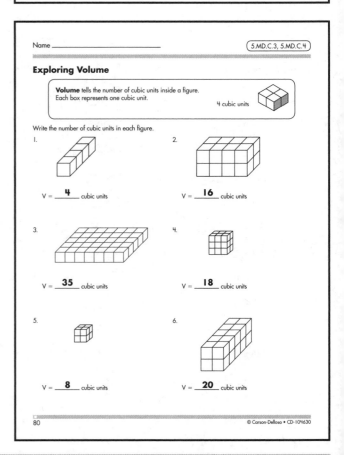

Answer Key

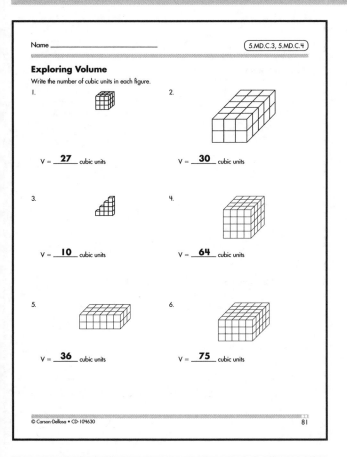

Name _____   5.MD.C.3, 5.MD.C.4

Exploring Volume
Write the number of cubic units in each figure.

1.

V = __**27**__ cubic units

2.

V = __**30**__ cubic units

3.

V = __**10**__ cubic units

4.

V = __**64**__ cubic units

5.

V = __**36**__ cubic units

6.

V = __**75**__ cubic units

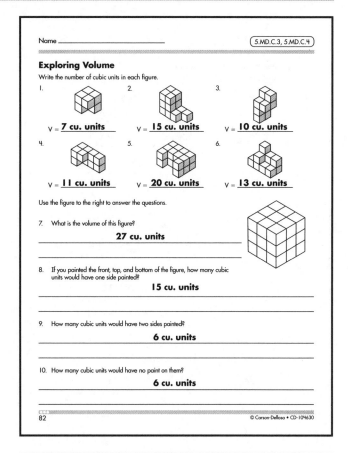

Name _____   5.MD.C.3, 5.MD.C.4

Exploring Volume
Write the number of cubic units in each figure.

1.   V = __**7 cu. units**__

2.   V = __**15 cu. units**__

3.   V = __**10 cu. units**__

4.   V = __**11 cu. units**__

5.   V = __**20 cu. units**__

6.   V = __**13 cu. units**__

Use the figure to the right to answer the questions.

7.   What is the volume of this figure?

27 cu. units

8.   If you painted the front, top, and bottom of the figure, how many cubic units would have one side painted?

15 cu. units

9.   How many cubic units would have two sides painted?

6 cu. units

10. How many cubic units would have no paint on them?

6 cu. units

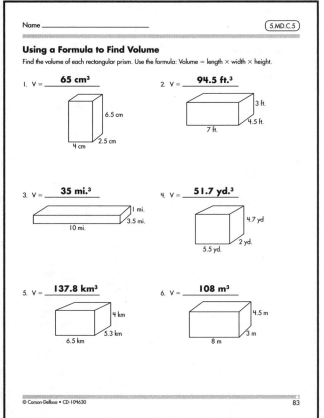

Name _____   5.MD.C.5

Using a Formula to Find Volume
Find the volume of each rectangular prism. Use the formula: Volume = length × width × height.

1.   V = __**65 cm³**__
     6.5 cm, 4 cm, 2.5 cm

2.   V = __**94.5 ft.³**__
     3 ft, 7 ft, 4.5 ft.

3.   V = __**35 mi.³**__
     1 mi., 10 mi., 3.5 mi.

4.   V = __**51.7 yd.³**__
     4.7 yd, 5.5 yd., 2 yd.

5.   V = __**137.8 km³**__
     4 km, 6.5 km, 5.3 km

6.   V = __**108 m³**__
     4.5 m, 8 m, 3 m

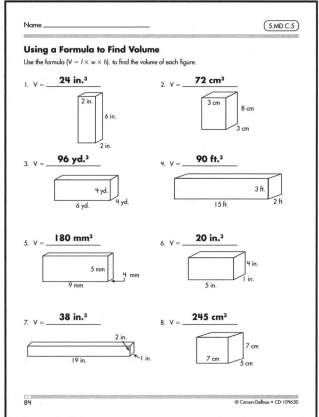

Name _____   5.MD.C.5

Using a Formula to Find Volume
Use the formula (V = l × w × h). to find the volume of each figure.

1.   V = __**24 in.³**__
     2 in., 6 in., 2 in.

2.   V = __**72 cm³**__
     3 cm, 8 cm, 3 cm

3.   V = __**96 yd.³**__
     4 yd., 6 yd., 4 yd.

4.   V = __**90 ft.³**__
     3 ft., 15 ft., 2 ft.

5.   V = __**180 mm³**__
     5 mm, 9 mm, 4 mm

6.   V = __**20 in.³**__
     4 in., 5 in., 1 in.

7.   V = __**38 in.³**__
     2 in., 19 in., 1 in.

8.   V = __**245 cm³**__
     7 cm, 7 cm, 5 cm

Answer Key

Name

5.MD.C.5

Using a Formula to Find Volume

Use the formula ($V = l \times w \times h$) to find the volume of each rectangular prism.

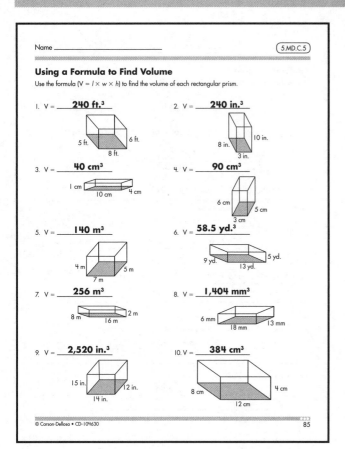

1. V = **240 ft.³**

2. V = **240 in.³**

3. V = **40 cm³**

4. V = **90 cm³**

5. V = **140 m³**

6. V = **58.5 yd.³**

7. V = **256 m³**

8. V = **1,404 mm³**

9. V = **2,520 in.³**

10. V = **384 cm³**

© Carson-Dellosa • CD-104630   85

Name

5.MD.C.5

Finding Volume

Use the formula to find volume: $V = l \times w \times h$

Find the volume of each figure.

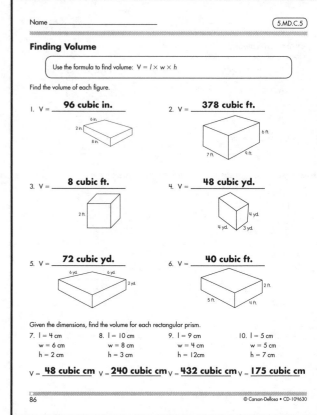

1. V = **96 cubic in.**

2. V = **378 cubic ft.**

3. V = **8 cubic ft.**

4. V = **48 cubic yd.**

5. V = **72 cubic yd.**

6. V = **40 cubic ft.**

Given the dimensions, find the volume for each rectangular prism.

7. l = 4 cm
w = 6 cm
h = 2 cm

8. l = 10 cm
w = 8 cm
h = 3 cm

9. l = 9 cm
w = 4 cm
h = 12cm

10. l = 5 cm
w = 5 cm
h = 7 cm

V = **48 cubic cm** V = **240 cubic cm** V = **432 cubic cm** V = **175 cubic cm**

86   © Carson-Dellosa • CD-104630

Name

5.MD.C.5

Finding Volume

Find the volume of each figure.

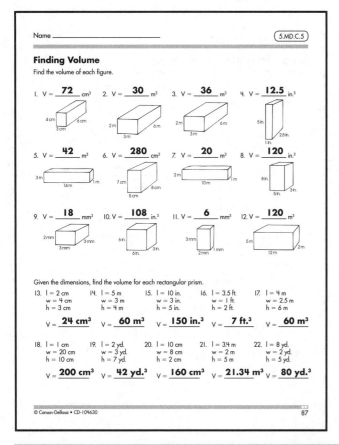

1. V = **72** cm³

2. V = **30** m³

3. V = **36** m³

4. V = **12.5** in.³

5. V = **42** m³

6. V = **280** cm³

7. V = **20** m³

8. V = **120** in.³

9. V = **18** mm³

10. V = **108** in.³

11. V = **6** mm³

12. V = **120** m³

Given the dimensions, find the volume for each rectangular prism.

13. l = 2 cm
w = 4 cm
h = 3 cm

14. l = 5 m
w = 3 m
h = 4 m

15. l = 10 in.
w = 1 ft.
h = 5 in.

16. l = 3.5 ft.
w = 1 ft.
h = 2 ft.

17. l = 4 m
w = 2.5 m
h = 6 m

V = **24 cm³** V = **60 m³** V = **150 in.³** V = **7 ft.³** V = **60 m³**

18. l = 1 cm
w = 20 cm
h = 10 cm

19. l = 2 yd.
w = 3 yd.
h = 7 yd.

20. l = 10 cm
w = 8 cm
h = 2 cm

21. l = 3.4 m
w = 2.5 m
h = 5 m

22. l = 8 yd.
w = 2 yd.
h = 5 yd.

V = **200 cm³** V = **42 yd.³** V = **160 cm³** V = **21.34 m³** V = **80 yd.³**

© Carson-Dellosa • CD-104630   87

Name

5.MD.C.5

Finding Volume

Find the volume of each figure.

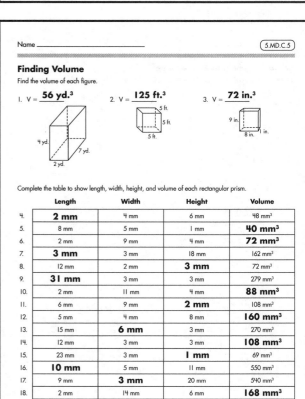

1. V = **56 yd.³**

2. V = **125 ft.³**

3. V = **72 in.³**

Complete the table to show length, width, height, and volume of each rectangular prism.

	Length	Width	Height	Volume
4.	**2 mm**	4 mm	6 mm	48 mm³
5.	8 mm	5 mm	1 mm	**40 mm³**
6.	2 mm	9 mm	4 mm	**72 mm³**
7.	**3 mm**	3 mm	18 mm	162 mm³
8.	12 mm	2 mm	**3 mm**	72 mm³
9.	**31 mm**	3 mm	3 mm	279 mm³
10.	2 mm	11 mm	4 mm	**88 mm³**
11.	6 mm	9 mm	**2 mm**	108 mm³
12.	5 mm	4 mm	8 mm	**160 mm³**
13.	15 mm	**6 mm**	3 mm	270 mm³
14.	12 mm	3 mm	3 mm	**108 mm³**
15.	23 mm	3 mm	**1 mm**	69 mm³
16.	**10 mm**	5 mm	11 mm	550 mm³
17.	9 mm	**3 mm**	20 mm	540 mm³
18.	2 mm	14 mm	6 mm	**168 mm³**
19.	**4 mm**	1 mm	15 mm	60 mm³

88   © Carson-Dellosa • CD-104630

Answer Key

Additive Volume

To find the **volume** of complex rectangular prisms, follow these steps:

1. First, break them into parts.
2. Then, find the volume of each part.
3. Finally, add the volume of the parts together.

Find the volume of each figure.

1.
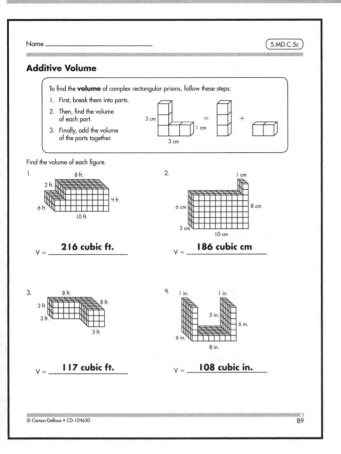

V = **216 cubic ft.**

2.

V = **186 cubic cm**

3.

V = **117 cubic ft.**

4.

V = **108 cubic in.**

Additive Volume

Remember, to find the volume of complex rectangular prisms, find the volume of each part, then add the parts together.

Find the volume of each figure.

1.
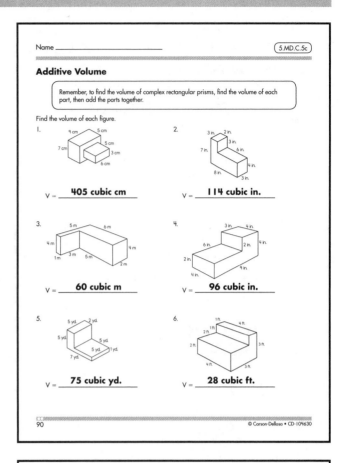

V = **405 cubic cm**

2.

V = **114 cubic in.**

3.

V = **60 cubic m**

4.

V = **96 cubic in.**

5.

V = **75 cubic yd.**

6.

V = **28 cubic ft.**

Additive Volume

Find the volume of each figure.

1.
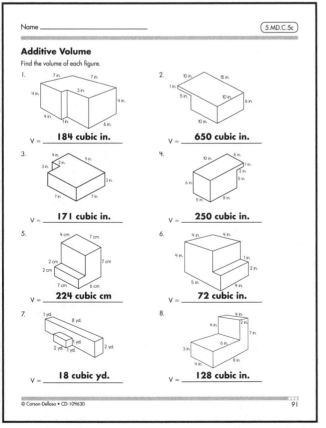

V = **184 cubic in.**

2.

V = **650 cubic in.**

3.

V = **171 cubic in.**

4.

V = **250 cubic in.**

5.

V = **224 cubic cm**

6.

V = **72 cubic in.**

7.

V = **18 cubic yd.**

8.

V = **128 cubic in.**

Volume Word Problems

Circle the correct answer for each problem.

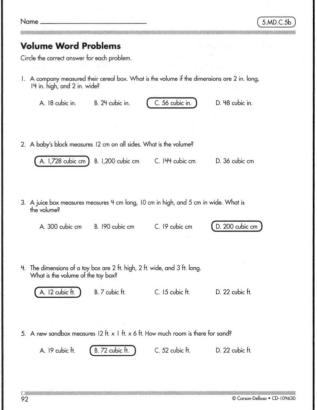

1. A company measured their cereal box. What is the volume if the dimensions are 2 in. long, 14 in. high, and 2 in. wide?

   A. 18 cubic in.   B. 24 cubic in.   **C. 56 cubic in.**   D. 48 cubic in.

2. A baby's block measures 12 cm on all sides. What is the volume?

   A. 1,728 cubic cm   B. 1,200 cubic cm   C. 144 cubic cm   D. 36 cubic cm

3. A juice box measures measures 4 cm long, 10 cm in high, and 5 cm in wide. What is the volume?

   A. 300 cubic cm   B. 190 cubic cm   C. 19 cubic cm   **D. 200 cubic cm**

4. The dimensions of a toy box are 2 ft. high, 2 ft. wide, and 3 ft. long. What is the volume of the toy box?

   A. 12 cubic ft.   B. 7 cubic ft.   C. 15 cubic ft.   D. 22 cubic ft.

5. A new sandbox measures 12 ft. x 1 ft. x 6 ft. How much room is there for sand?

   A. 19 cubic ft.   **B. 72 cubic ft.**   C. 52 cubic ft.   D. 22 cubic ft.

Answer Key

Name _____ 5.MD.C.5b

Volume Word Problems

Answer each question.

1. Monty and his friend are building a wooden frame for a garden. They want it to measure 10 feet long, 5 feet wide, and 1 foot high. What will the volume of the frame be?
50 cubic ft.

2. The shed in the James's yard is 16 yd. long, 4 yd. wide, and 5 yd. high. What is the space inside the shed?
320 cubic yd.

3. Your neighbors are pouring concrete for their driveway's foundation. The foundation will be 0.25 feet deep and 24 feet long by 12 feet wide. How much concrete will they need to complete the job?
72 cubic ft.

4. A shipping box is 18 inches long, 12 inches wide, and 6 inches tall. What volume of products can it hold?
1,296 cubic in.

5. A bag is 14 inches tall, 10 inches wide, and 4 inches deep. Is it large enough to hold Natalie's school books, which have a combined volume of 500 cubic inches? If so, how much space will she have left? If not, how much more space does she need?
Yes. She will have 60 cubic in. of space left.

Name _____ 5.MD.C.5b, 5.MD.C.5c

Volume Word Problems

Answer each question.

1. Luke is making a 3-layer square cake. Each layer has a side length of 25 centimeters, and a depth of 7 cm. What volume of cake batter will he need to make the cake?
13,125 cubic cm of cake batter

2. A student desk is 20 inches wide, 18 inches deep, and 6 inches tall. If each of Mischa's 4 books takes up 144 cubic inches, how much empty space will Mischa have in her desk?
1,584 cubic in.

3. Pilar digs a hole in the garden that is 2 feet wide, 4 feet long, and 1 foot deep. Her brother Hector digs a hole 4 feet square and 3 feet deep. What volume of dirt did they remove altogether?
56 cubic ft.

4. A ball pit is 6 meters long, 8 meters wide, and 2 meters deep. If about 85 balls fit in 1 cubic meter, about how many balls will it take to fill the ball pit?
about 8,160 balls

5. The swimming pool is 5 yards wide, 9 yards long, and 5 feet deep. If the pool fills at a rate of 100 cubic feet an hour, how long will the pool take to fill?
$20\frac{1}{4}$ hours, or 20 hours and 15 minutes

Name _____ 5.G.A.1

Graphing Coordinates

Coordinates are like directions for placing a point on a coordinate plane.

(3, 4)

- Always start at 0.
- The first number, 3, tells you how many spaces to move over, or along the x-axis.
- The second number, 4, tells you how many spaces to move up, or along the y-axis.

So, for (3, 4) you should move over 3 and up 4 to locate the point.

Graph and label each pair of coordinates.

1. A (3, 4)
2. B (1, 8)
3. C (5, 1)
4. D (3, 7)
5. E (8, 2)

Graph and label each pair of coordinates.

6. F (2, 9)
7. G (10, 7)
8. H (6, 9)
9. I (1, 5)
10. J (4, 3)

Name _____ 5.G.A.1

Graphing Coordinates

Remember, in a coordinate pair the first number tells you how many spaces to move over. The second number tells you how many spaces to move up.

Graph and label each pair of coordinates.

1. A (3, 5)
2. B (7, 8)
3. C (1, 3)
4. D (6, 10)
5. E (9, 4)
6. F (8,1)

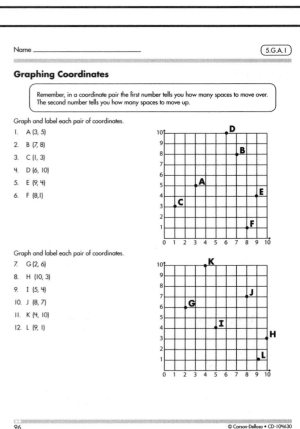

Graph and label each pair of coordinates.

7. G (2, 6)
8. H (10, 3)
9. I (5, 4)
10. J (8, 7)
11. K (4, 10)
12. L (9, 1)

Answer Key

Graphing Coordinates

Graph and label each pair of coordinates.

1. A (6, 5)
2. B (1, 7)
3. C (10, 9)
4. D (8, 3)
5. E (5, 0)
6. F (7, 2)
7. G (0, 5)
8. H (2, 9)

Graph and label each pair of coordinates.

9. I (9, 5)
10. J (0, 8)
11. K (1, 1)
12. L (6, 4)
13. M (2, 0)
14. N (7, 2)
15. O (7, 0)
16. P (3, 3)

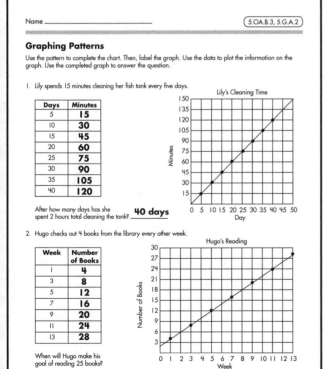

Graphing Patterns

Use the pattern to complete the chart. Use the data to plot the information on the graph. Use the completed graph to answer the question.

1. Nina is making necklaces for her friends. Each necklace uses 25 beads.

Necklaces	Number of Beads
1	25
2	50
3	**75**
4	**100**
5	**125**
6	**150**
7	**175**
8	**200**

Beads come in packs of 100.
How many necklaces can Nina make with one pack? **4 necklaces**

2. Davis earns $6 each week for doing extra chores at home.

Week	Amount
1	$6
2	$12
3	**$18**
4	**$24**
5	**$30**
6	**$36**
7	**$42**
8	**$48**

How much money does Davis earn after 6 weeks? **$36**

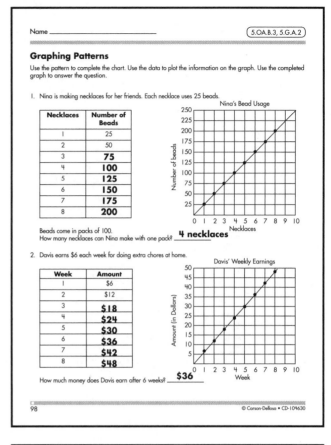

Graphing Patterns

Use the pattern to complete the chart. Then, label the graph. Use the data to plot the information on the graph. Use the completed graph to answer the question.

1. Lily spends 15 minutes cleaning her fish tank every five days.

Days	Minutes
5	**15**
10	**30**
15	**45**
20	**60**
25	**75**
30	**90**
35	**105**
40	**120**

After how many days has she spent 2 hours total cleaning the tank? **40 days**

2. Hugo checks out 4 books from the library every other week.

Week	Number of Books
1	**4**
3	**8**
5	**12**
7	**16**
9	**20**
11	**24**
13	**28**

When will Hugo make his goal of reading 25 books?

in between week 11 and week 13

Graphing Patterns

Use the pattern to complete the chart. Then, label the graph. Use the data to plot the information on the graph. Use the completed graph to answer the question.

1. Xander spends 20 minutes practicing his guitar every other day.

Day	Time (in min.)
1	**20**
3	**40**
5	**60**
7	**80**
9	**100**
11	**120**
13	**140**
15	**160**

How many days does it take him to practice 3 hours total? **17 days**

2. Ana runs 1.5 laps at soccer practice each week.

Week	Number of Laps
1	**1.5**
2	**3**
3	**4.5**
4	**6**
5	**7.5**
6	**9**

How many laps does Ana run every month (every 4 weeks)? **6 laps**

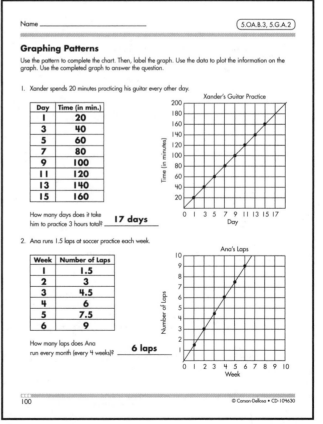

Answer Key

Understanding Attributes of Two-Dimensional Figures

Attributes of a shape are the characteristics that define it and separate it from similar shapes. Number of sides and angles, and types of angles and lines are all examples of attributes.

Polygons

equilateral triangle	parallelogram	rectangle	regular hexagon
rhombus	right triangle	square	trapezoid

List the shapes that share each attribute.

1. 4 sides
**parallelogram
rectangle
rhombus
square
trapezoid**

2. equal sides
**equilateral triangle
rhombus
square**

3. equal, opposite sides
**parallelogram
rectangle
regular hexagon
rhombus
square**

4. at least one pair of parallel sides
**parallelogram
rectangle
regular hexagon
rhombus
square
trapezoid**

5. 4 angles
**parallelogram
rectangle
rhombus
square
trapezoid**

6. at least 1 right angle
**rectangle
right triangle
square**

7. 4 right angles
**rectangle
square**

8. equal, opposite angles
**parallelogram
rectangle
regular hexagon
rhombus
square**

Understanding Attributes of Two-Dimensional Figures

For each shape, write its attributes.

1. rectangle
**4 sides
2 pairs of equal, parallel sides
4 right angles**

2. square
**4 equal sides
2 pairs of parallel sides
4 right angles**

3. trapezoid
**4 sides
2 opposite, parallel sides**

4. rhombus
**4 equal sides
2 pairs of opposite, parallel sides
2 pairs of equal, opposite angles**

5. parallelogram
**4 sides
2 pairs of equal, parallel sides
2 pairs of equal, opposite angles**

6. quadrilateral
4 sides

7. Can a parallelogram be a rectangle? Why or why not?
Yes, a parallelogram only needs equal, opposite sides and angles. If the angles are all right angles, it is also a rectangle.

8. What makes a rhombus different from a square?
A rhombus does not have to have right angles.

9. What attribute makes a trapezoid different from other quadrilaterals?
It only needs 1 pair of parallel sides.

Understanding Attributes of Two-Dimensional Figures

equilateral triangle	parallelogram	rectangle	rhombus
right triangle	square	trapezoid	

Use the polygon names to answer each question. Each word may be used more than once or not at all.

1. A square is also a **rectangle** because it has 4 right angles and equal, opposite sides.

2. A **parallelogram** can not be a rectangle unless it has 4 right angles.

3. A **rhombus** is only a square if it has 4 right angles.

4. A **trapezoid** will never be a rectangle because it only has one pair of equal, opposite sides.

5. A **equilateral triangle**, **square**, and a **rhombus** all have equal sides.

6. An **equilateral triangle** is the only 3-sided figure that can have a right angle.

7. A **trapezoid** is the only quadrilateral that can have only 1 right angle.

8. A **rhombus** is like a **parallelogram** because they are quadrilaterals with equal and opposite angles that don't have to be right angles.

Congratulations!

_____

_____

receives this award for

Signed _____ Date _____

$$\begin{array}{r} 2{,}563 \\ \times\ \ 41 \\ \hline \end{array}$$

$$\begin{array}{r} 1{,}359 \\ \times\ \ 15 \\ \hline \end{array}$$

$$\begin{array}{r} 658 \\ \times\ 986 \\ \hline \end{array}$$

$$6\overline{)5{,}470}$$

$$\begin{array}{r} 251 \\ \times\ 741 \\ \hline \end{array}$$

$$\begin{array}{r} 459 \\ \times\ \ 28 \\ \hline \end{array}$$

$$\begin{array}{r} 159 \\ \times\ \ 36 \\ \hline \end{array}$$

$$25\overline{)3{,}455}$$

$$\begin{array}{r} 554 \\ \times\ \ 14 \\ \hline \end{array}$$

$$\begin{array}{r} 258 \\ \times\ 133 \\ \hline \end{array}$$

$$\begin{array}{r} 5{,}627 \\ \times\ \ \ 42 \\ \hline \end{array}$$

$$5\overline{)655}$$

$$\begin{array}{r} 522 \\ \times\ 326 \\ \hline \end{array}$$

$$\begin{array}{r} 111 \\ \times\ \ 85 \\ \hline \end{array}$$

$$\begin{array}{r} 212 \\ \times\ \ 12 \\ \hline \end{array}$$

$$12\overline{)3{,}600}$$

105,083   185,991   7,756   170,172

20,385   12,852   34,314   9,435

648,788   5,724   236,334   2,544

911r4   138r5   131   300

Change to a mixed number.

$$\frac{45}{10}$$

Change to a mixed number.

$$\frac{62}{11}$$

Change to a mixed number.

$$\frac{20}{17}$$

Change to a mixed number.

$$\frac{125}{41}$$

Change to an improper fraction.

$$7\frac{1}{4}$$

Change to an improper fraction.

$$15\frac{1}{2}$$

Change to an improper fraction.

$$3\frac{6}{15}$$

Change to an improper fraction.

$$12\frac{3}{8}$$

Find the equivalent.

$$\frac{5}{25} = \frac{}{50}$$

Find the equivalent.

$$\frac{5}{8} = \frac{}{48}$$

Find the equivalent.

$$\frac{6}{20} = \frac{}{100}$$

Find the equivalent.

$$\frac{4}{9} = \frac{}{63}$$

$$\frac{2}{9} \times \frac{2}{9} =$$

$$\frac{1}{4} \times \frac{7}{8} =$$

$$2 \times \frac{5}{8} =$$

$$4 \times \frac{4}{5} =$$

$$3\frac{2}{41} \qquad 1\frac{3}{17} \qquad 5\frac{7}{11} \qquad 4\frac{1}{2}$$

$$\frac{99}{8} \qquad \frac{51}{15} \qquad \frac{31}{2} \qquad \frac{29}{4}$$

$$28 \qquad 30 \qquad 30 \qquad 10$$

$$3\frac{1}{5} \qquad 1\frac{1}{4} \qquad \frac{7}{32} \qquad \frac{4}{81}$$

$$3\frac{1}{3} \times 1\frac{7}{9} =$$

© CD

$$4 \div \frac{1}{2} =$$

© CD

$$\frac{1}{3} \div 9 =$$

© CD

$$\frac{1}{2} \div 12 =$$

© CD

$$2\frac{6}{11} \times 5\frac{1}{2} =$$

© CD

$$10 \div \frac{3}{4} =$$

© CD

$$\frac{6}{7} \div 4 =$$

© CD

$$\frac{3}{8} \div 3 =$$

© CD

$$\frac{1}{8} \times \frac{1}{7} =$$

© CD

$$6 \div \frac{2}{3} =$$

© CD

$$3 \div \frac{5}{8} =$$

© CD

$$\frac{3}{4} \div 4 =$$

© CD

$$\frac{3}{7} \times \frac{1}{10} =$$

© CD

$$8 \div \frac{1}{4} =$$

© CD

$$5 \div \frac{3}{5} =$$

© CD

$$\frac{2}{5} \div 10 =$$

© CD

$5\frac{25}{27}$

8

$\frac{1}{27}$

$\frac{1}{24}$

14

$13\frac{1}{3}$

$\frac{3}{14}$

$\frac{1}{8}$

$\frac{1}{56}$

9

$4\frac{4}{5}$

$\frac{3}{16}$

$\frac{3}{70}$

32

$8\frac{1}{3}$

$\frac{1}{25}$

$$\frac{2}{12} + \frac{3}{24}$$

$$\frac{1}{6} + \frac{1}{4}$$

$$\frac{11}{12} - \frac{1}{6}$$

$$8\frac{1}{2} - 8\frac{1}{4}$$

$$\frac{2}{4} + \frac{5}{7}$$

$$\frac{3}{5} + \frac{4}{7}$$

$$11\frac{1}{3} - 3\frac{2}{3}$$

$$19 - \frac{1}{2}$$

$$5\frac{1}{6} + 7\frac{8}{9}$$

$$2\frac{3}{8} + 5\frac{1}{2}$$

$$4\frac{1}{2} + 7\frac{3}{10}$$

$$8 - \frac{5}{8}$$

$$6\frac{2}{5} + 4\frac{2}{3}$$

$$4\frac{7}{8} + 6\frac{3}{4}$$

$$5\frac{2}{5} + 4\frac{3}{8}$$

$$\frac{5}{7} - \frac{2}{9}$$

$$\frac{7}{24} \qquad 1\frac{3}{14} \qquad 13\frac{1}{18} \qquad 11\frac{1}{15}$$

$$\frac{5}{12} \qquad 1\frac{6}{35} \qquad 7\frac{7}{8} \qquad 11\frac{5}{8}$$

$$\frac{3}{4} \qquad 7\frac{2}{3} \qquad 11\frac{4}{5} \qquad 9\frac{31}{40}$$

$$\frac{1}{4} \qquad 18\frac{1}{2} \qquad 7\frac{3}{8} \qquad \frac{31}{63}$$

$$\begin{array}{r} 35.6 \\ +\ 77.7 \\ \hline \end{array}$$

$$\begin{array}{r} 0.3 \\ +\ 0.05 \\ \hline \end{array}$$

$$\begin{array}{r} 44.2 \\ +\ 445 \\ \hline \end{array}$$

$$\begin{array}{r} 14.6 \\ +\ 12.18 \\ \hline \end{array}$$

$$\begin{array}{r} 8.5 \\ -\ 3.2 \\ \hline \end{array}$$

$$\begin{array}{r} 122.52 \\ -\ 65.3 \\ \hline \end{array}$$

$$\begin{array}{r} 34.9 \\ -\ 13.3 \\ \hline \end{array}$$

$$\begin{array}{r} 45.4 \\ -\ 5.4 \\ \hline \end{array}$$

$$\begin{array}{r} 47.8 \\ \times\ 1.23 \\ \hline \end{array}$$

$$\begin{array}{r} 41.2 \\ \times\ 1.1 \\ \hline \end{array}$$

$$\begin{array}{r} 5.2 \\ \times\ 0.16 \\ \hline \end{array}$$

$$\begin{array}{r} 15.46 \\ \times\ 0.4 \\ \hline \end{array}$$

$0.2\overline{)16}$

$0.9\overline{)10.8}$

$8\overline{)0.48}$

$6\overline{)2.4}$

© CD

113.3   0.35   489.2   26.78

5.3   57.22   21.6   40

58.794   45.32   0.832   6.184

80   12   0.06   0.4

Find the volume.

$l = 9$ cm
$w = 9$ cm
$h = 10$ cm

Find the volume.

$l = 7$ mm
$w = 1$ mm
$h = 8$ mm

Find the volume.

$l = 2$ in.
$w = 5$ in.
$h = 3$ in.

Find the volume.

$l = 22$ mm
$w = 15$ mm
$h = 14$ mm

Find the volume.

$l = 13$ in.
$w = 12.75$ in.
$h = 16$ in.

Find the volume.

$l = 5$ ft.
$w = 4$ ft.
$h = 2$ ft.

Find the volume.

$l = 1.5$ yd.
$w = 1.5$ yd.
$h = 2$ yd.

Find the volume.

$l = 8.5$ cm
$w = 6.3$ cm
$h = 6.5$ cm

Find the volume.

$l = 10.4$ ft.
$w = 13.8$ ft.
$h = 4.4$ ft.

Find the volume.

$l = 5.5$ in.
$w = 25.8$ in.
$h = 35.2$ in.

Find the volume.

$l = 50$ in.
$w = 46.5$ in.
$h = 33.3$ in.

Find the volume.

$l = 10$ cm
$w = 18$ cm
$h = 58$ cm

Find the volume.

4 cm
3 cm
2 cm

Find the volume.

4 mm
6 mm
3 mm

Find the volume.

$1\frac{1}{2}$ ft.
$2\frac{2}{3}$ ft.
$3\frac{2}{3}$ ft.

Find the volume.

8.75 in.
5 in.
4.2 in.

30 in.^{3} 40 ft.^{3} 56 mm^3 810 cm^3

348.075 cm^3 4.5 yd.^{3} 2652 in.^{3} 4620 mm^3

10440 cm^3 77422.5 in.^{3} 4994.88 in.^{3} 631.488 ft.^{3}

183.75 in.^{3} $14\frac{2}{3}$ ft.^{3} 72 mm^3 24 cm^3